NIST Technical Note 1483

Measurements of Heat and Combustion Products in Reduced-Scale Ventilation-Limited Compartment Fires

Matthew Bundy
Anthony Hamins
Erik L. Johnsson
Sung Chan Kim
Gwon Hyun Ko
David B. Lenhert

U.S. Department of Commerce
Technology Administration
Building and Fire Research Laboratory
National Institute of Standards
and Technology
Gaithersburg, MD 20899

July 2007

U.S Department of Commerce
Carlos M. Gutierrez, *Secretary*
Technology Administration
Robert Cresanti, *Under Secretary of Commerce for Technology*
National Institute of Standards and Technology
William Jeffery, *Director*

Disclaimer:
Certain commercial equipment, instruments, or materials are identified in this document. Such identification does not imply recommendation or endorsement by the National Institute of Standards and Technology, nor does it imply that the products identified are necessarily the best available for the purpose.

Table of Contents

iv

List of Figures

vii

List of Tables

1 Introduction

A series of new reduced-scale compartment fire experiments were conducted, which included local measurements of temperature and species composition. The measurements are unique to the compartment fire literature. By design, the experiments provided a comprehensive and quantitative assessment of major and minor carbonaceous gaseous species and soot at two locations in the upper layer of fire in a 2/5 scale International Organization for Standards (ISO) 9705 room. The enclosure defined in the international standard ISO 9705 "Full-scale room test for surface products" [1] is an important structure in which to conduct fire research. Many dozens of research projects and journal articles have focused on this enclosure and the standard describing its use. It is a common reference point for studies of many fire-related phenomena as well as fire modeling efforts.

While some previous studies have considered the mixture fraction to analyze experimental compartment fire data, few have considered minor hydrocarbon species and none have considered soot. In tandem, accurate measurements of temperature at these same locations allowed analysis of thermal effects on species concentrations. A wide range of fuel types were considered, including aliphatic hydrocarbons (natural gas and heptane), aromatic hydrocarbons (toluene and polystyrene) and alcohols (methanol and ethanol).

Field models, such as the National Institute of Standards and Technology (NIST) Fire Dynamics Simulator (FDS) [2], are widely used by fire protection engineers to predict fire growth and smoke transport for practical engineering applications. Field models numerically solve the conservation equations of mass, momentum and energy that govern low-speed, thermally-driven flows with an emphasis on smoke and heat transport from fires. All field models have strengths and weaknesses. Among the various assumptions used in the development of previous versions of FDS, all chemical species were tied to the mixture fraction state relations. A single mixture fraction variable cannot be used for the prediction of carbon monoxide and soot, and the yield of these species was prescribed in FDS 4, rather than predicted. In fact, the yield of these species is usually not constant, but a complex function of their time-temperature history. In practice, an engineer using FDS 4 would choose combustion product yields directly from literature values for well-ventilated burning, using data from a bench-scale apparatus [3]. Using this approach, the carbon monoxide (CO) volume fraction for pool fire burning in an under-ventilated compartment can be underestimated by as much as a factor of ten. A new version of FDS (version 5) is currently being tested which implements a predictive model of CO production.

The experimental results provided in this report are the first step of a long-term NIST project to generate the data necessary to test our understanding of fire phenomena in enclosures and to guide the development and validation of field models by providing high quality experimental data. The experimental plan was designed in cooperation with developers of the NIST FDS model to assure that the measurements would be of maximum value. Advanced development of FDS and other field models is extremely important, since it will lead to improved accuracy in the prediction of underventilated burning, typical of fire conditions that occur in structures. Improving models for under-ventilated burning will foster improved prediction of important life safety and fire dynamic phenomena, including fire spread, backdraft, flashover, and egress (involving the presence of toxic gases and smoke), which are critically important for application of fire models for fire safety. In summary, the main objective of this project is to provide an improved understanding of the physics, chemistry, and structure of underventilated compartment fires, and to provide experimental measurements to guide the development of fire chemistry sub-models.

1.1 Background and Relationship to Current Research

Experimental research on enclosure fires has been on-going in fire research laboratories and academic institutions over the last 50 years. The motivation has varied from applied investigations studying particular fire scenarios to more fundamental work with the goal of understanding toxic species production behavior in fires. Some of the fundamental research that tried to ascertain ventilation and upper-layer effects on enclosure fire chemistry was conducted in well-controlled hoods. Sometimes, the main objectives of this research was to generally develop and validate fire models or particular structural fire simulations, while much of the research was conducted to acquire a better understanding of complex enclosure fire dynamics with a focus on chemical and thermal conditions. This section provides an overview of some of the recent research efforts in enclosure fires and highlights some of the more pertinent experimental work.

Research conducted at Harvard University and the California Institute of Technology in the 1980s explored fires burning under an exhaust hood (false ceiling) to simulate the layer effect of an enclosure fire, e.g. [4,5]. The relative distance of the fire below the hood was adjusted to vary the entrainment of air into the plume before it entered the upper layer. These experiments focused on underventilated burning, pathways for air to enter the upper layer, and the validity of the concept of "global equivalence ratio" (GER) which is the fuel-to-air mass ratio normalized by the mass ratio required for stoichiometric burning. Some recent modeling work by Cleary and Kent [6], has also focused on experimental data from hoods. In a recent study, Brohez et al. explored the use of a bench-scale calorimeter to measure fire properties of materials burning in underventilated conditions [7,8].

Research at NIST by Bryner et al. further explored the global equivalence ratio concept and carbon monoxide production in a reduced (2/5) scale enclosure with natural gas as the principal fuel [9]. The results showed that the upper layer in enclosure fires is not homogeneous, and that CO can be produced in greater quantities than predicted by the GER concept, depending on temperatures and flow patterns developed within an enclosure. The current effort is meant to overlap some of the conditions explored by Bryner et al. and to repeat and fill gaps in the data. Pitts expanded the work to full-scale and other fuels such as heptane and wood. It was established that wood pyrolysis in the upper layer of an enclosure fire can produce high concentrations of CO directly without further oxidation to CO_2 [10]. A subsequent study by Lattimer confirmed and expanded on this research [11].

Researchers at Virginia Tech investigated fires in a reduced-scale enclosure that directed the air inflow through slots in the floor connected to a duct where instrumentation was used to quantify air entrainment [12]. Several fuels were studied, and this configuration produced results consistent with GER predictions due to the more distinct, less dynamic nature of the gas layer structure. Later work used a more typical enclosure design and focused on transport of gas species outside the doorway and how it was affected by doorway geometry, soffit design, and hallway configuration [13]. More recently, Gann et al [14] conducted research on transport of toxic species in a full-scale enclosure with a corridor. These data were analyzed by Hirschler [15]. Researchers in Sweden conducted a study [16] of underventilated fires in an ISO 9705 room with a window vent of varying height. Several polymer fuel types were included in this study and measurements of local equivalence ratio and toxic gas species were performed.

Pitts [17] provides a comprehensive review of the application of the GER concept to predict CO concentration in building fires, using data from the Harvard and Cal Tech hood experiments [3,4], the Virginia Tech enclosure studies [11], and the NIST reduced-scale enclosure experiments [8,9]. Several CO formation mechanisms were identified, which were substantiated by detailed chemical kinetic modeling. While the GER concept is of limited utility for predicting the local CO concentration, important aspects of enclosure fire dynamics and chemistry are highlighted in this paper.

Several recent experimental studies [18,19,20] have used very small scale enclosures (0.21 m^3, 0.06 m^3, and 0.05 m^3, respectively) while investigating underventilated burning of propane and heptane fires. These bench-scale studies described the structure and dynamics of underventilated burning including extinction, flame projection and flame stability. Another recent study [21,22] has used an intermediate-scale enclosure similar to that used for this paper, but a roof vent was added as well.

Recently, NIST has conducted a number of high-profile case studies in which realistic-scale mock-ups of actual fire scenarios were recreated with the ultimate goal of improving building codes and standards. These studies included the World Trade Center disaster investigation [23], the Rhode Island Station nightclub fire [24], and the Chicago Cook County Administration Building fire investigation [25]. The compartment fires in all of these studies burned real furnishings and became underventilated as the fire evolved. In addition, a series of large-scale compartment fire experiments were conducted to simulate an over-ventilated fire in a nuclear power plant cable room [26] to provide data for fire model validation.

1.2 Approach and Scope

The series of experiments reported on here was conducted in a reduced scale (2/5 ISO 9705 room) enclosure (RSE). The experiments repeated and extended a part of the work of Bryner and coworkers [9]. Similar to Bryner's experiments, natural gas served as a fuel; the burning of heptane, toluene, methanol, ethanol, and polystyrene was also investigated. In most experiments, the fuel was controlled and metered by flow valves or pumped into a pool burner or spray nozzle. Experiments were run to near-steady conditions. Multiple fire sizes were run consecutively to decrease the time required to approach steady-state. Ventilation was varied during some experiments by modifying the door opening. Two types of enclosure lining materials were investigated and compared.

Temperature and species composition measurements were made at many of the same nominal locations as studied previously by Bryner [9]. Measurements included CO, CO_2, temperature, heat fluxes, and dynamic pressures (used to obtain velocities). One emphasis of this series was to develop techniques for the measurement of hydrocarbons and soot. Hydrocarbons were measured with Flame Ionization Detector (FID) total hydrocarbon analyzers and gas chromatography (GC). The GC measurements were used to independently validate the total hydrocarbon measurements and to allow accurate determination of carbon mass distribution. The quantification of hydrocarbon species was needed to describe the chemical structure of underventilated fires. Soot samples were extracted from within the enclosure and measured gravimetrically. Optical soot measurements were performed at the doorway.

The fuels included in this test series were selected to cover a wide range of combustion properties and to simulate fuels encountered in actual building fires. Gases, liquids and solids were selected for testing to cover a wide range of physical properties. Realistic materials represent complex multi-component fuels. In this study, all of the fuels selected were homogeneous single component fuels to simplify the analysis and attempt to find generalizable trends in the results. Real materials are often oxygenated. This includes many types of commodity materials including nylon (e.g., carpet), cellulose (e.g. paper and building products), polyester (e.g., fabric), epoxy (e.g., adhesives), polymethylmethacrylate (PMMA), and POM (polyoxometalate). In this study, alcohols were selected as a surrogate to represent the compartment fire chemistry in the burning of oxygenated fuels.

In a real compartment fire, fuel sources are physically distributed throughout the compartment. In this study, a single location for the fuel was used to simplify the analysis of the experimental results for the purpose of model validation. Multiple fuel locations would have led to uncertainty in the specification of the location of the heat source, which is a critical boundary condition in a CFD fire calculation and, therefore, crucial information for model validation.

In a real compartment fire, heat feedback and natural ventilation give rise to important aspects of the structure and dynamics of the fire, such as the temperature field and the spatial distribution of combustion products. This study deliberately set out to investigate representative fire conditions at two key locations in the upper layer of the compartment, which were selected based on a series of CFD fire modeling calculations. The upper layer locations were selected to provide two distinct conditions in the upper layer, one relatively close to the natural ventilation flow of fresh air through the doorway and the other relatively far from the doorway, on the far side of the fire source. The design calculations confirmed that these locations would provide a range of local conditions in terms of the combustion species equivalence ratio and the temperature that would be useful for the construction of a database for model development and validation. To enhance the range of conditions investigated and in an attempt to seek information on the relationship between the combustion products and generalizable local flame conditions, a broad range of fire heat release rates and a number of very different fuel types were selected for study. At the same time, the effect of compartment ventilation was changed to induce a range of mixing and compartment fire conditions.

2 Experimental Method

2.1 Reduced Scale Enclosure

2.1.1 Design and construction

Experiments were conducted using an enclosure, shown in Fig. 1, that is roughly a 2/5 scale replicate of the ISO 9705 room [1]. The steel frame for this enclosure was used extensively in the early 1990s to study carbon monoxide production in compartment fires. A detailed description of the design and construction of the Reduced Scale Enclosure (RSE) can be found in the original NIST report [9]. The original report described the "as designed" internal dimensions of the enclosure as 98 cm wide × 98 cm tall × 146 cm deep, however for this report, the "as constructed" internal dimensions were measured as 95 cm wide × 98 cm tall × 142 cm deep, with a pre-burn uncertainty of less than 1 cm on each dimension. The uncertainty in the internal dimensions increased as more fire experiments were conducted, e.g., during and after some of the fire tests, the walls were observed to deform in local areas by as much as 10 cm.

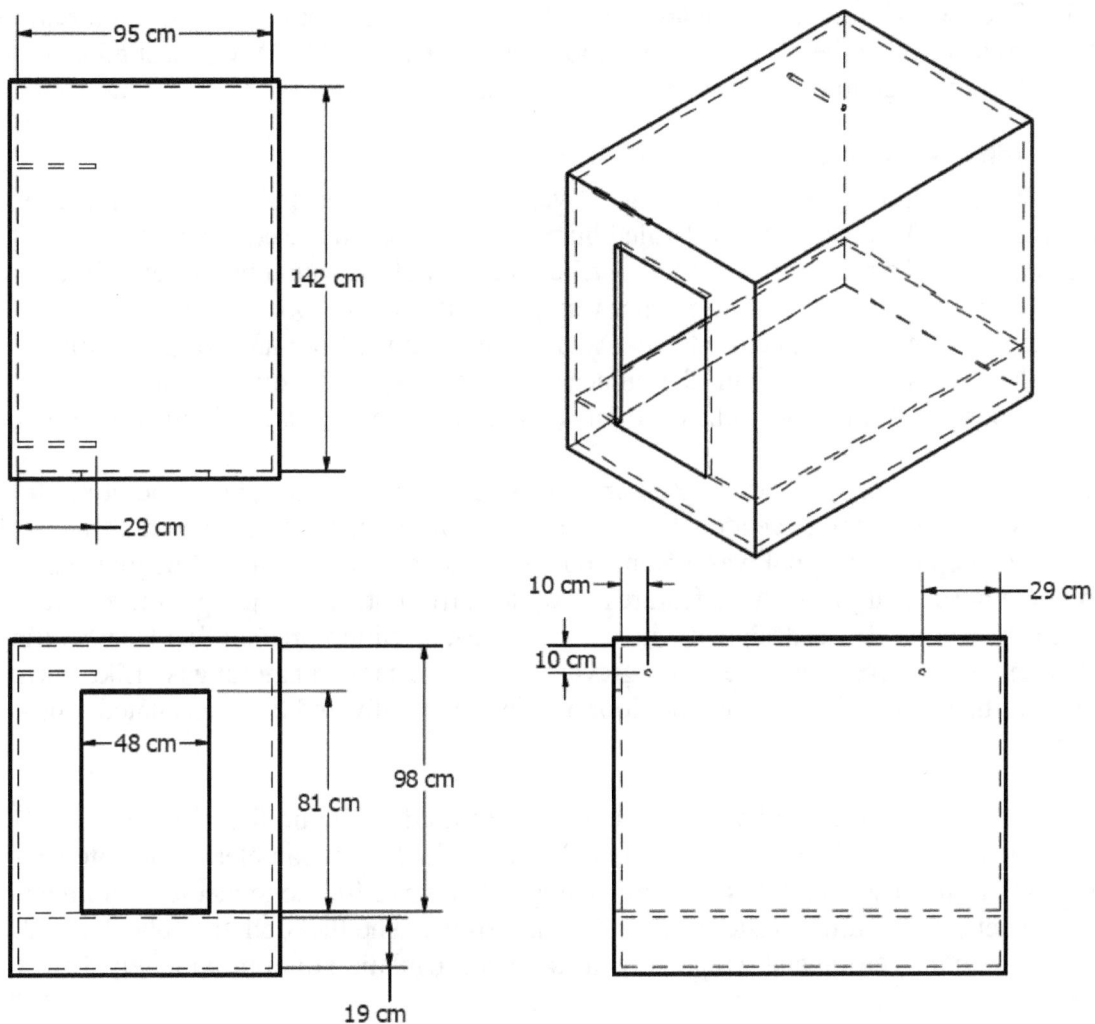

Figure 1. Perspective views of the Reduced Scale Enclosure and upper layer gas sampling probes drawn to scale. Dimensions are given with respect to interior walls.

The steel frame of the enclosure was lined with 2 layers of 1.27 cm thick insulation board. For the first six tests, a calcium silicate board (Marinite I) was used. For all other tests a rigid self supporting ceramic fiber (alumina and silica) board (Kaowool M-board) was used. The location of the retention bolts for the Marinite board were the same as the original test series [9]. Because the M-board sheet size was 122 cm × 91 cm, a single board would not span the length of the enclosure and joint seams were present. Stainless steel furnace pins were used to secure the M-board in both the original bolt locations (six retention points per wall and ceiling) and near the seams. The performance of the two different lining materials is discussed in Sec. 4.3. A comparison of the fire tests using the different lining materials showed no significant effect on the gas temperature and species measurements.

2.1.2 Doorway Variations
The standard doorway geometry (shown in Fig. 1) was 81 cm tall × 48 cm wide and centered horizontally on the 95 cm front wall. The bottom of the door was aligned with the inside floor. The inside floor was 43 cm above the laboratory floor. This configuration was used for all but two of the tests described here. The narrow door tests (listed in Table 1) were test #5 and test #6. The narrow doorway geometry was 81 cm tall × 24 cm wide.

2.2 Burner Designs
Four different burner designs, shown in Fig. 2, were used in this test series to accommodate the different fuels. A 13 cm square gravel-filled burner (Burner A) was used for the first three tests using natural gas. The area of this burner matched the area of the round burner used in the original test series [9]. The rim of the burner was 15 cm above the floor. Natural gas was delivered to the burner by a 1.3 cm tube that was fed through the floor and wrapped with Kaowool blanket insulation. The insulation nearly filled the space below the burner. The square geometry of burners A and B was chosen to match the rectangular grid used in FDS simulations.

A 25 cm square liquid cooled burner (Burner B) was used for both natural gas and liquid fuels. The burner was designed to have a pool surface area that increased with the depth of the pool. The maximum depth of the pool was 6.5 cm and the burner walls were at a 24 degree angle with respect to the horizontal plane. This feature allows for different size steady pool fires with a single burner. Burner B was designed with a 2.5 cm vertical rim to prevent fuel from spilling out of the burner. This burner was filled with gravel for some tests with natural gas. Like burner A, the height of the rim was 15 cm from the floor and fuel was delivered by an insulated tube through the floor.

A water-cooled downward spray burner (Burner C) was used for liquid fuels in tests #11, #12 and #15. The nozzle was located 20 cm above the base of a 40 cm diameter round catch pan with a 12.5 cm rim. The spray was delivered using a 90 degree full-cone medium atomization (droplet diameter ≈ 250 μm) nozzle with a 1.40 mm orifice. The fuel delivery tube was fed through a hole in the ceiling and wrapped with Kaowool insulation of approximately 3 cm thickness.

Polystyrene pellets were burned using round pans 22 cm, 40 cm and 60 cm in diameter (Burners D, E and F respectively). Each of the burners was centered on the floor. The pan size was increased for this fuel in order to reach under-ventilated conditions. A description of the test conditions including burner type can be found in Table 1.

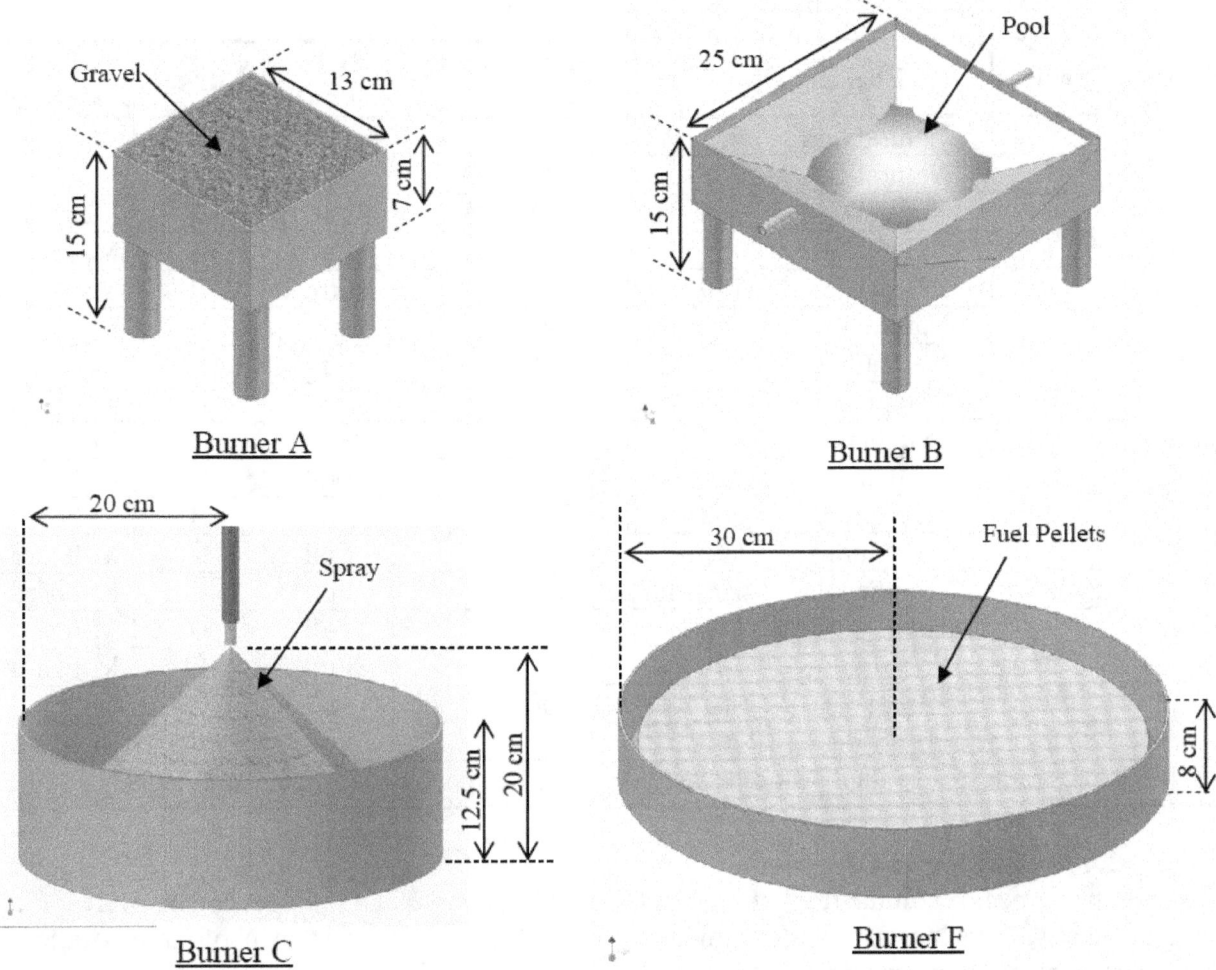

Figure 2. Dimensional drawing of burners used in the RSE experiments. Burners D and E were similar to Burner F, but the diameters were 22 cm and 40 cm, respectively.

2.3 Experimental Conditions

Experiments were conducted during two separate series. The test number (#), series and controlled test parameters are listed in Table 1. The two main differences between the series 1 and series 2 were the wall lining material (see Sec. 2.1.1) and the gas sample conditioning systems (see Sec. 2.5.3).

The fuels included in this test series are listed in Table 1 and included gases, liquids and solids at ambient temperature. The composition of natural gas used for these tests is described in Sec. 3.1. The heptane fuel was a blend of heptane isomers. The fuel referred to as ethanol was actually a blend of 90 % ethanol and 10 % methanol by volume. The polystyrene fuel was clear granulated (2.5 g / 100 granules) Dow Styron 666D general purpose resin with a manufacturer reported average molecular mass of 230.8 kg/mol.

Table 1. List of test numbers and key experimental conditions.

Test #	Series	Fuel	Heat Release Rates* (kW)	Door Vent	Burner	Wall Material
1	1	Natural Gas	75 ,190, 75	Full	A	Marinite I
2	1	Natural Gas	255, 395, 180, 115, 50	Full	A	Marinite I
3	1	Natural Gas	265, 410, 180, 115, 75	Full	A	Marinite I
4	1	Heptane	155, 270, 375	Full	B	Marinite I
5	1	Heptane	140, 220	Narrow	B	Marinite I
6	1	Natural Gas	75, 175, 270, 420, 80	Narrow	B	Marinite I
6.5	2	Natural Gas	95, 425, 270, 180, 85	Full	B	M board
7	2	Heptane	150, 245, 340	Full	B	M-board
8	2	Methanol	15	Full	B	M-board
9	2	Ethanol	20	Full	B	M-board
10	2	Toluene	50, 140, 200, 295, 340	Full	B	M-board
11	2	Ethanol	80, 145, 265, 335	Full	C	M-board
12	2	Methanol	70, 140, 240, 305	Full	C	M-board
13	2	Polystyrene	15	Full	D	M-board
14	2	Polystyrene	70	Full	E	M-board
15	2	Heptane	90, 160, 225, 300, 375, 85	Full	C	M-board
16	2	Polystyrene	360, 310	Full	F	M-board

* Nominal pseudo steady state heat release rate values from calorimetry measurements

2.4 Measurement Locations

Temperature, species volume fraction, soot mass fraction and velocity measurements were conducted at various locations in the compartment doorway and interior. A photograph of the front gas, soot, and temperature measurement probes is shown in Fig. 3. The sample probe for the gravimetric soot measurement is seen on the right of the image, and the aspirated thermocouple protrudes down through the ceiling. Figure 4 and Fig. 5 show the relative positions (drawn to scale) of the measurement probes in the doorway and inside the enclosure respectively. The reference point used to describe the positions within the enclosure is annotated in Fig. 5.

The measurement locations inside the RSE are listed in Table 2 and the locations in the doorway are listed in Table 3. The column heading (data label) corresponding to these measurements locations are also listed in the measurement location tables. These data labels are referenced in the tables and figures in Section 3. A complete list of data column headings can be found in Appendix C.

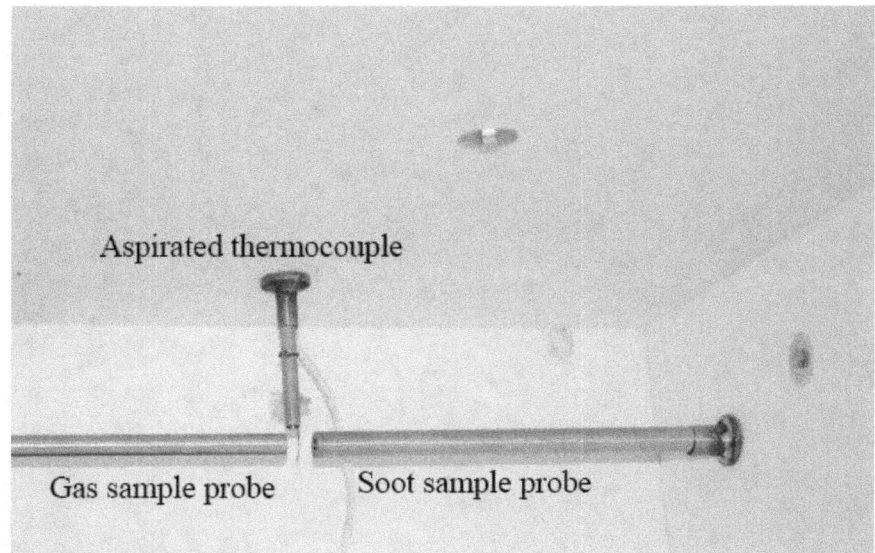

Figure 3. Photograph of extractive sampling probes and aspirated thermocouple at the front sample location (photo taken with rear wall removed).

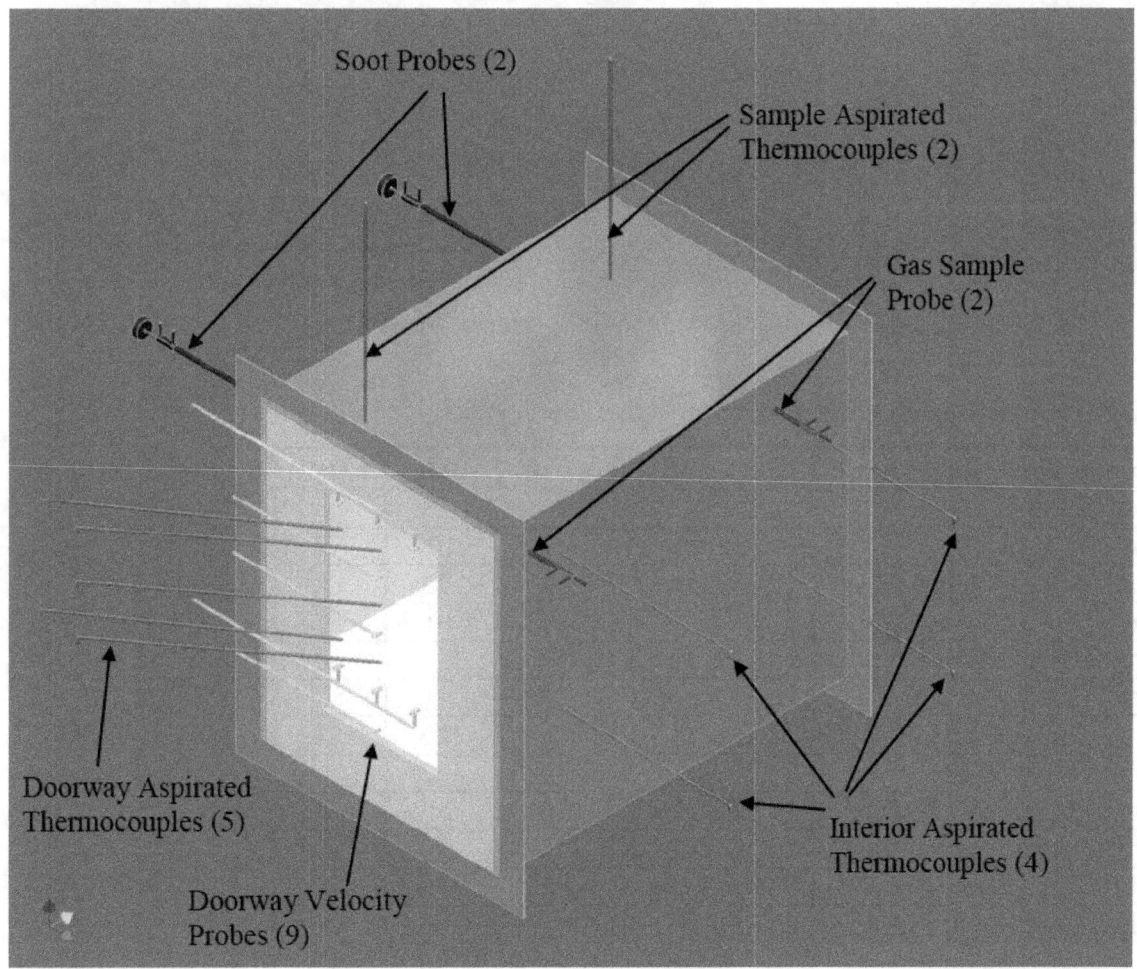

Figure 4. Isometric view of Reduced Scale Enclosure showing the relative position of doorway measurement probes.

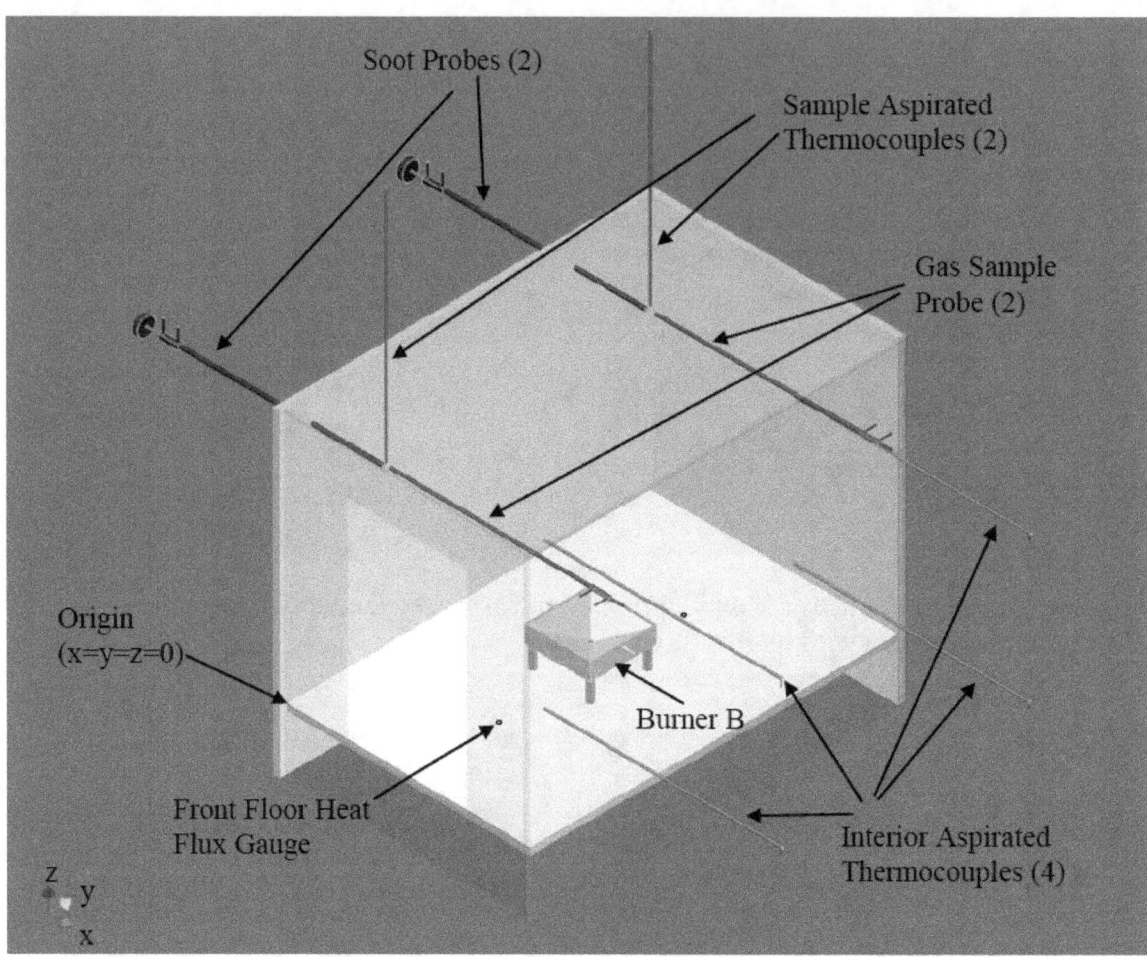

Figure 5. Isometric semi-transparent view of the RSE interior measurement probe locations and burner.

Table 2. Location of measurement probes inside of the enclosure.

Probe Description	Data Label	x (cm)	y (cm)	z (cm)
Gas Sample Rear	1 O2Rear	29	113	88
Gas Sample Front	7 O2Front	29	10	88
Gravimetric Soot Sample Rear	6 SootRear	29	113	88
Gravimetric Soot Sample Front	11 SootRear	29	10	88
Aspirated Thermocouple	15 TRSampA	29	113	88
Aspirated Thermocouple	16 TFSampA	29	10	88
Aspirated Thermocouple	17 TR24A	75	122	24
Aspirated Thermocouple	18 TR80A	75	122	80
Aspirated Thermocouple	19 TF24A	75	20	24
Aspirated Thermocouple	20 TF80A	75	20	80
Total Heat Flux Gauge Rear	12 HFR	48	106	0
Total Heat Flux Gauge Front	13 HFF	48	35	0
Bare Bead Thermocouple	21 TFloorR	49	106	0
Bare Bead Thermocouple	22 TFloorF	49	35	0
Bare Bead Thermocouple	23 TCeilF	31	11	98

Table 3. Location of measurement probes in the enclosure doorway.

Probe Description	Data Label	x (cm)	y (cm)	z (cm)
Aspirated Thermocouple	31 TC70CA	48	-5	70
Aspirated Thermocouple	29 TC70LA	32	-5	70
Aspirated Thermocouple	36 TC50CA	48	-5	50
Aspirated Thermocouple	39 TC30CA	48	-5	30
Aspirated Thermocouple	33 TC30LA	32	-5	30
Bi-Directional Velocity Probe	45 VD79L	32	-5	79
Bi-Directional Velocity Probe	46 VD79C	48	-5	79
Bi-Directional Velocity Probe	47 VD79R	64	-5	79
Bi-Directional Velocity Probe	48 VD60C	48	-5	60
Bi-Directional Velocity Probe	49 VD40C	48	-5	40
Bi-Directional Velocity Probe	50 VD20L	32	-5	20
Bi-Directional Velocity Probe	51 VD20C	48	-5	20
Bi-Directional Velocity Probe	52 VD20R	64	-5	20
Bi-Directional Velocity Probe	53 VD5C	48	-5	5

2.5 Measurement Instrumentation
2.5.1 Heat Release Rate

Heat Release Rate (HRR) measurements were conducted using the 3 m × 3 m calorimeter at the NIST Large Fire Research Laboratory (LFRL). The HRR measurement was based on the oxygen consumption calorimetry principle first proposed by Huggett [27]. This method assumes that a known amount of heat is released for each gram of oxygen consumed by a fire. The measurement of exhaust flow velocity and gas volume fractions (O_2, CO_2 and CO) were used to determine the HRR based on the formulation derived by Parker [28]. A detailed description of the methodology used for this measurement can be found in a previous report [29]. The experimental apparatus for the current measurements has been modified since the earlier report was written. In 2005, the 3 m × 3m square hood was installed in the LFRL. A schematic drawing of the 3 m hood is shown in Fig. 6. The exhaust flow rate, optical soot and extractive gas measurements were performed in a vertical section of the 48.3 cm diameter duct. A bi-directional probe located 9 cm from the edge of the duct was used to measure the exhaust flow velocity. Because of the non-uniform shape of the velocity profile, a flow calibration coefficient was used in the HRR calculation. The flow coefficient was determined using natural gas calibration performed before and after the test series. The flow calibration coefficients $\pm 2\sigma$ for these tests ranged from 0.85 ± 0.04 to 0.90 ± 0.05.

The exhaust gas was sampled through a perforated tube across the duct downstream of the velocity probe. Figure 7 shows the exhaust gas sampling system. The main difference between this system and the one previously reported [29] is the method for removing water from the gas sample. The current system uses a Nafion dryer instead of a dry ice cold trap. Nafion is a copolymer of tetrafluoroethylene (Teflon) and perfluoro-3,6-dioxa-4-methyl-7-octene-sulfonic acid. A dew point meter was added to monitor the efficiency of the gas dryer. The dew point temperature meter measures the change in electrical impedance of a hygroscopic conductive polymer in the range of -80 °C to 20 °C. The delay time from the gas sample tube to the analyzers was 20 s. Measurements of exhaust soot and total hydrocarbons were performed, but

11

were not included in the HRR calculation because in most cases they have negligible effect. The combined expanded relative uncertainty of the HRR measurements reported here was 14 %, based on a propagation of uncertainty analysis [29]. The exhaust mass flow rate was the largest component of uncertainty in the HRR measurement. A list of commercial equipment used for all of the measurements described in this report can be found in Appendix D.

2.5.2 Fuel Metering

Two different fuel delivery systems were used to control and measure the flow rate of fuel to the burners. The natural gas tests used a positive displacement flow meter with a standard relative uncertainty of 1 % to measure the fuel volume flow rate. Combined with measurements of the fuel temperature, pressure, and ideal heat of combustion, the ideal natural gas burner heat release rate was determined with a combined expanded uncertainty of 2.4 %. A gas chromatograph was used to measure the composition of the natural gas [30]. The net ideal heating value of the natural gas was determined using the composition measurements [31].

The liquid fuel delivery rate was measured using a dual rotor turbine flow meter with a manufacturer's stated uncertainty of 0.1 % in the range from 0.06 L/min to 11 L/min. Although the liquid fuel volume flow rate was accurately measured, the fuel mass burning rate was not directly measured. In some cases, the amount of fuel (depth of liquid pool) in the burners was observed to vary with time, even though the fuel delivery rate was constant. Once the fuel delivery was stopped, the burnout of existing fuel could take several minutes.

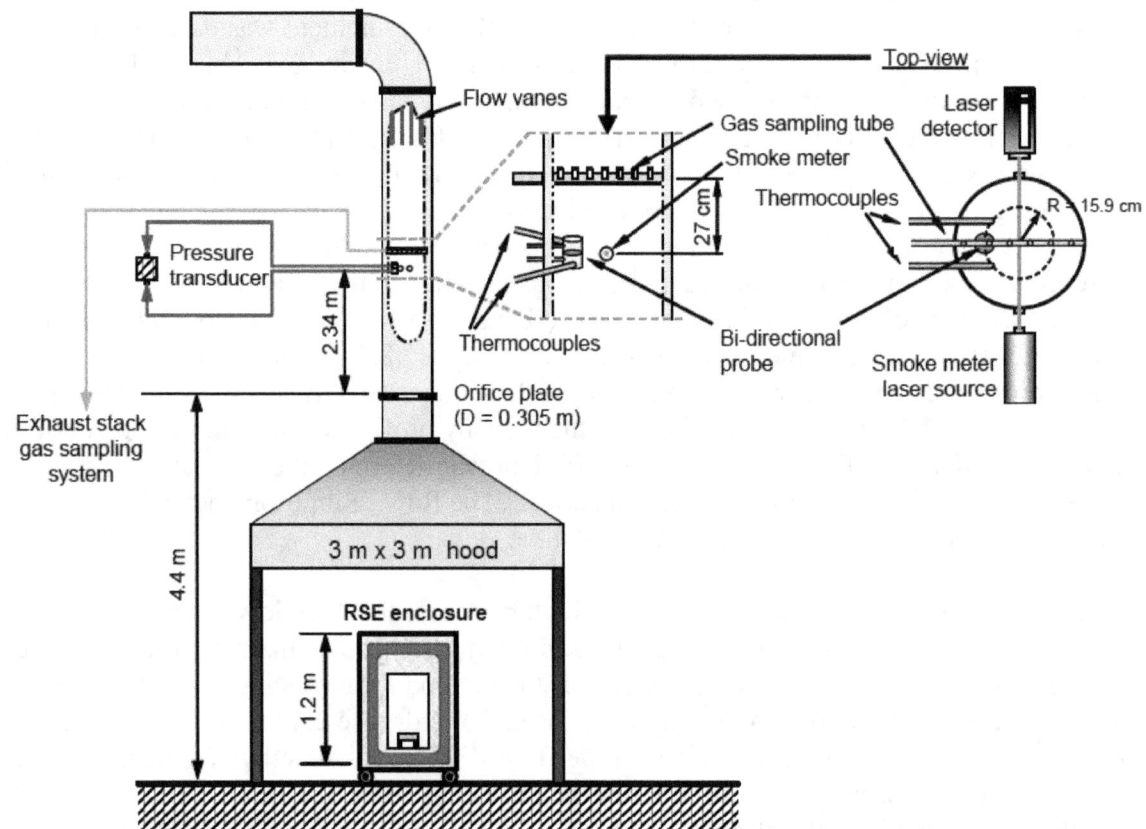

Figure 6. Schematic drawing of 3 m hood and exhaust stack instrumented for calorimetry and light extinction measurements.

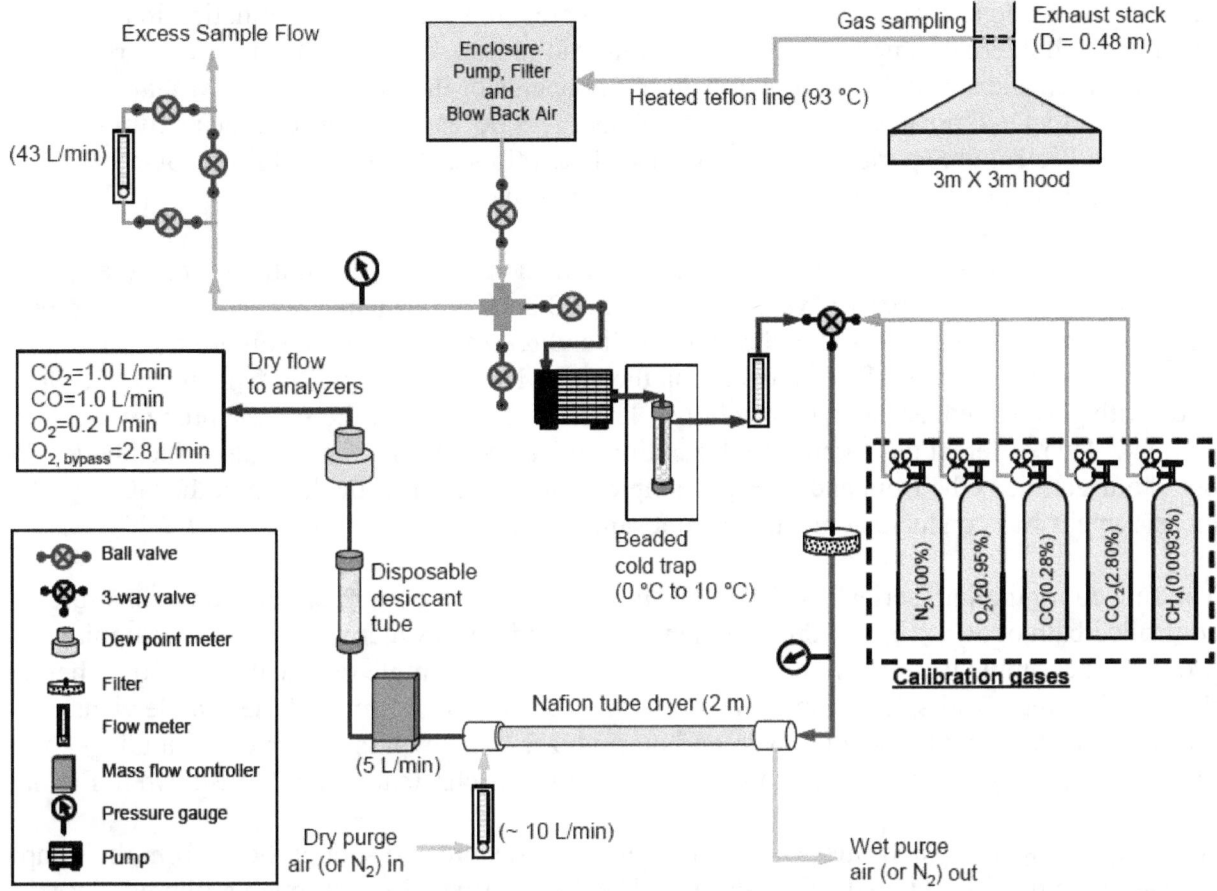

Figure 7. Exhaust gas sampling system used for heat release rate measurement.

2.5.3 Gas Species

Gas species were continuously measured at two locations (front and rear) inside the RSE during each of the tests. Oxygen was measured using paramagnetic analyzers. The 10 % to 90 % response time (t_{10-90}) of the oxygen analyzer was less than 12 s. Carbon monoxide and carbon dioxide were measured using non-dispersive infrared (NDIR) analyzers. The t_{10-90} response time for the CO_2/CO analyzers was less than 5 s. Total hydrocarbons were measured using two flame ionization detectors (FID) having a t_{10-90} response time of less than 1 s.. A gas chromatograph (GC) was used intermittently during some of the tests at the front gas sampling location. The cycle time on the GC measurements was 20 min to 30 min. The dried sample gas dew point temperature was measured using a thin polymer sensor. Soot and temperature were also measured at these two locations (see Sec. 2.5.4 and Sec. 2.5.5).

The two total hydrocarbon analyzers used in these experiments were designed to measure high volume fractions of hydrocarbons. The analyzers were factory calibrated for up to 50 % volume fraction of hydrocarbons as methane and were capable of measuring even higher concentrations. The primary span gas used for these tests was 20 % volume fractions of methane with a balance of nitrogen. A span gas of 1 % methane was also used to periodically check the linearity of the detector. The FID burner fuel used was 40 % hydrogen and 60 % nitrogen on a volumetric basis.

Each hydrocarbon analyzer had an internal filter to prevent soot from accumulating in the plumbing and internal sample pump which could lead to less sensitivity due to hydrocarbon contamination and also deterioration of some components of the instrument. It was later determined that additional external filtration of soot was necessary to protect the analyzer and enable a sufficient time period for sampling soot-laden flows. The external filter could be replaced much more frequently and easily than the internal filter.

Two liquid cooled probes were used to sample gas inside the enclosure at the front and rear locations. The 1 m long probes were constructed of 3 concentric stainless steel (type 304) tubes. Liquid coolant was forced through the inner shell and returned through the outer shell. This design allowed the cooling fluid to condition the entire length of the probe. The inner tube was lined with glass to reduce catalytic reactions. The inner diameter of the sample probe was 4.0 mm. Two different gas sample configurations were used during the tests described here. For both configurations the front and rear gas sample systems were identical, except the GC measurement was conducted only at the front sample location.

The first configuration (series 1 in Table 1), shown in Fig. 8, used a re-circulated temperature controlled bath of 50 % (by volume) ethylene glycol and 50 % water to cool the gas sample probes. The sample was drawn through a 3 m stainless steel sample line heated to 120 °C before the sample stream was split. Immediately after the heated line, 3 L/min of the sample went through a heated filter to the total hydrocarbon analyzer. The bypass stream of the total hydrocarbon analyzer was connected to the inlet of the gas chromatograph (not shown in Fig. 8).

A two-stage water trap and filter was used to remove moisture and soot particles from the sample path going to the O_2, CO_2 and CO analyzers. The sample first passed through a filtered glass trap cooled in a wet ice bath and then passed through a beaded glass trap cooled with dry ice. The sample pump was located downstream of the drying traps. The gas analyzers were connected in parallel so the total flow rate to the rack gas was 2.3 L/min. A 5 way ball valve was connected to each analyzer to switch between the gas sample, zero calibration gas and span calibration gas. The zero and span gas volume fractions are shown in Fig. 8. A dew point transducer was connected to the sample gas line prior to the oxygen analyzer. The oxygen analyzer had separate inlet ports for zero and span gases. The expanded (k = 2) relative uncertainty of each of the span gas volume fractions was ± 1 %.

A number of problems were encountered with the first sampling configuration. The coolant bath for the gas probes could not maintain a steady probe temperature. A vapor lock was formed in one of the probes and the loss of cooling caused the probe to melt and fail. The glass traps became clogged with ice after a period of time (creating a loss of sample flow) and there were intermittent leaks into the sample line through the trap seals. The internal filter in the total hydrocarbon analyzer became clogged with soot during some of the tests. After several tests, the gas sampling system was redesigned for improved performance. The redesigned gas sampling system (series 2 in Table 1) is shown in Fig. 9.

14

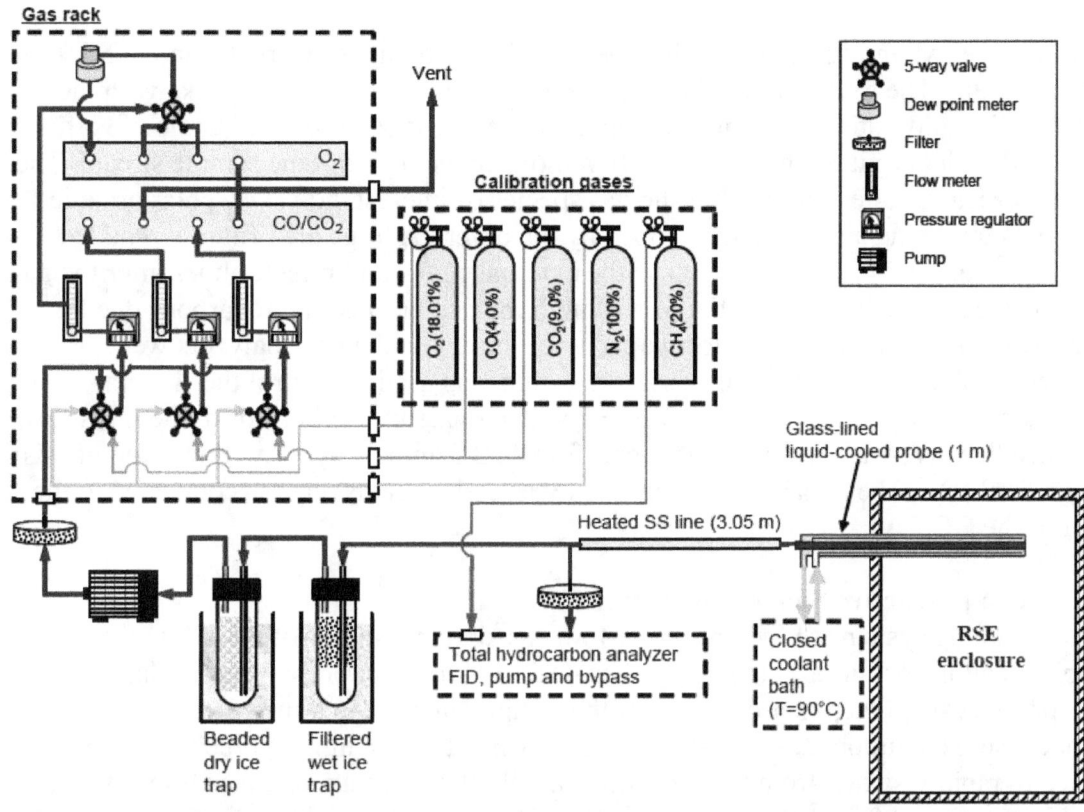

Figure 8. Schematic drawing of gas sampling system for series 1 tests #1 to test #6.

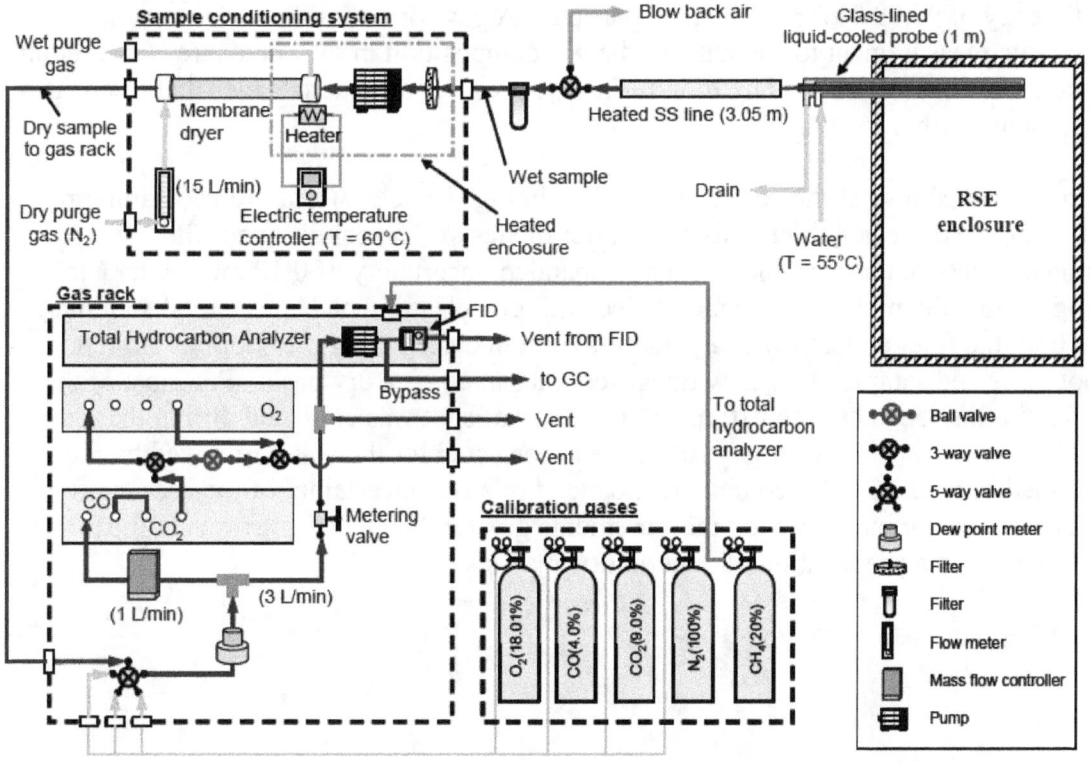

Figure 9. Schematic drawing of gas sampling system for series 2 (test #6.5 to test #16).

The sample probes shown in Figs. 3-5, 8, and 9 were cooled using house water heated to 55 °C at a flow rate of 1 L/min. The total hydrocarbon analyzers were placed in the gas racks with the other analyzers. The cold traps were replaced with a membrane type dryer. A bundle of Nafion tubes were purged with dry nitrogen to selectively remove moisture from the sample stream. The Nafion conditioner has no effect on most of the gas species of interest, however, polar organic compounds (i.e. ketones and alcohols) are trapped by the dryer. A large area filter was added between the heated line and gas dryer. Because the external filters and transfer lines after the gas dryer were not heated, there was a potential loss of high molecular mass hydrocarbons due to condensation. Due to limitations in the flow capacity of the dryer, the gas analyzers were connected in series. A mass flow controller set to 1 L/min was used to control the flow through the $O_2/CO_2/CO$ analyzers. The flow to the hydrocarbon analyzer was split prior to the mass flow controller. A needle valve was used to set the total flow to 3 L/min (only a small fraction of this passed through the FID). The bypass flow from the hydrocarbon analyzer was connected to the injection port of the GC.

2.5.4 Extractive Soot Measurement

A gravimetric sampling system (shown in Fig. 10) was used to measure soot mass fractions at the two sample locations within the enclosure. The design of the soot probe was similar to the gas sampling probes except the inner diameter of the sample tube was 6.4 mm. The soot sampling probes were conditioned with 65 °C water flowing at 1.0 L/min. A three way solenoid valve was used to rapidly switch from the bypass to sample flow. A sample gas mass flow rate of 2.75 standard L/min (N_2 @ 0 °C, 101.3 kPa) was drawn through the collection filter for a period of 60 s to 300 s. The collection filter was a 47 mm round Zeflour membrane filter with an aerosol retention efficiency of 99.99 % for 2 μm sized particles. A gas correction factor was applied to the mass flow rate measurement to account for the gas composition in the enclosure. The amount of time for sampling was determined by monitoring the pressure drop across the filter to ensure an optimal amount of filter loading.

The collection filters (shown at the base of the probes in Fig. 10 below) and probe cleaning pads were conditioned in a desiccant drier before and after the tests. The conditioned filters were weighed using an analytic mass balance with an expanded uncertainty of 0.12 mg. After each soot sampling period, the probe was cleaned twice with gun cleaning pads. The total soot mass collected on both the filter and 2 cleaning pads was used in determining the soot mass fraction. Both the soot mass and sample mass flow rates were measured on a dry basis. For most of the tests conducted in this series between 10 mg and 200 mg of soot was collected during the 1 min to 5 min sample time. The extracted gas volume was corrected for the water removed by the method described in Sec. 2.7. The combined expanded relative uncertainty of the soot mass fraction measurement (for mass fraction measurements greater than 0.001 g/g) was in the range of 2 % to 5 % based on a propagation of uncertainty analysis.

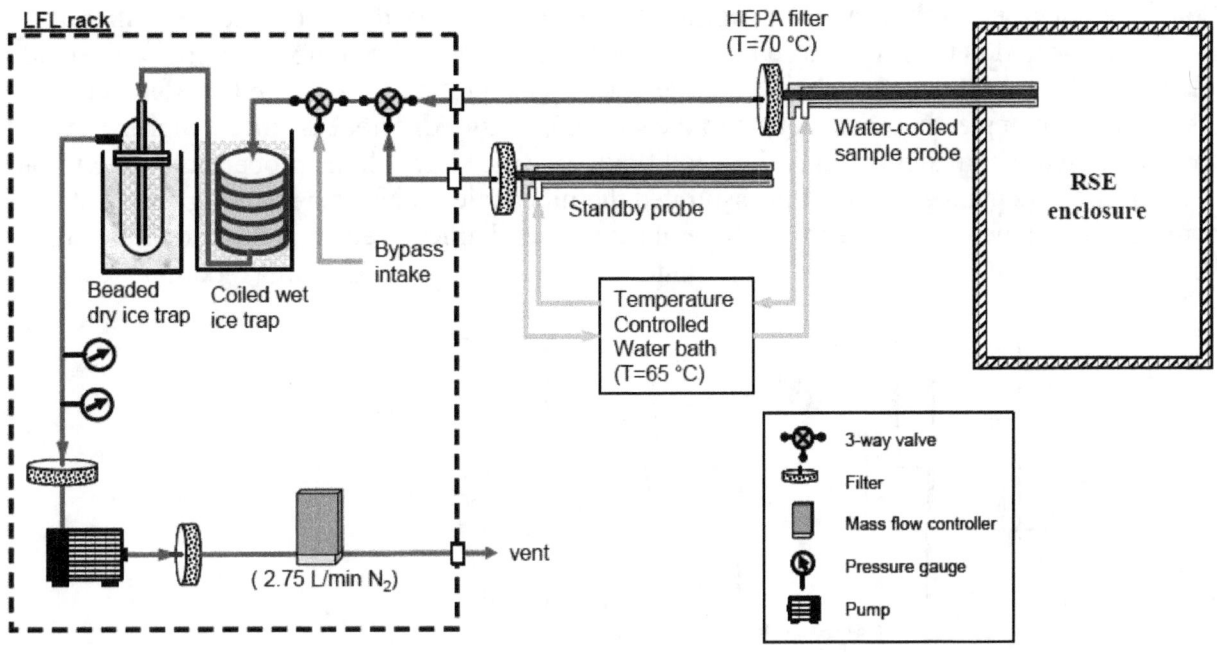

Figure 10. Schematic drawing of soot sampling system.

2.5.5 Temperatures

Aspirated thermocouples

A bare-bead thermocouple situated in a compartment fire typically experiences radiative exchange with walls, hot smoke, flames, and the surrounding environment with the effect that the measured temperature is not the true gas temperature. Accurate correction for these effects is complex, due to temporally and spatially varying local temperatures, velocities, and species. To reduce the effect of energy exchange on temperature measurement accuracy, aspirated thermocouple probes were used in addition to bare-bead thermocouples in this study.

An aspirated thermocouple probe is a bare-bead thermocouple contained within a small cylindrical metal tube through which the sample gas flows. If the flow over the bead is at least 5 m/s, a more accurate gas temperature measurement may be obtained [32]. According to Blevins [33], higher flows may be required depending on the thermal environment. Aspirated thermocouple probes may be shielded by a single cylindrical tube or by two or more concentric cylindrical tubes. In either case, the flow and thermal conditions and the detailed design of the assembly can impact measurement accuracy. Double-shielded aspirated thermocouple probes based on a design from National Advisory Committee for Aeronautics (NACA) were used in this study [34]. Figure 11 shows a drawing of an end-hole type NACA design aspirated thermocouple probe. Models with the entrance hole perpendicular to the probe axis were also used.

Each aspirated thermocouple was connected to a set of wet-ice and dry-ice traps, a flowmeter, and a pump using 9.5 mm (3/8 in) outer diameter (OD) copper and polyethylene tubing. A schematic of this is shown in Fig. 12. The gas was filtered and dried with the traps to protect the flowmeters and pumps. Flows were set at 24 L/min for each aspirated probe. While the volumetric flows were set at the flow meters to be the same for all probes, since the ice traps cooled the hot gases, high temperature compartment gases produced much higher velocities at the bead compared to those produced by low temperature gases. The uniform setting of the cold volumetric flows kept the mass flows consistent across the probes. This velocity difference effect was not completely proportionate to the gas temperature differences since a higher flow would experience a greater pressure drop and flow resistance through the probe and tubing. Due to the large flows pumped through the aspirated thermocouple probes, the resulting temperature represents a volumetric average over a several centimeter diameter region at the end of the probe. For further discussion of the probe and gas interaction see Appendix B of this report.

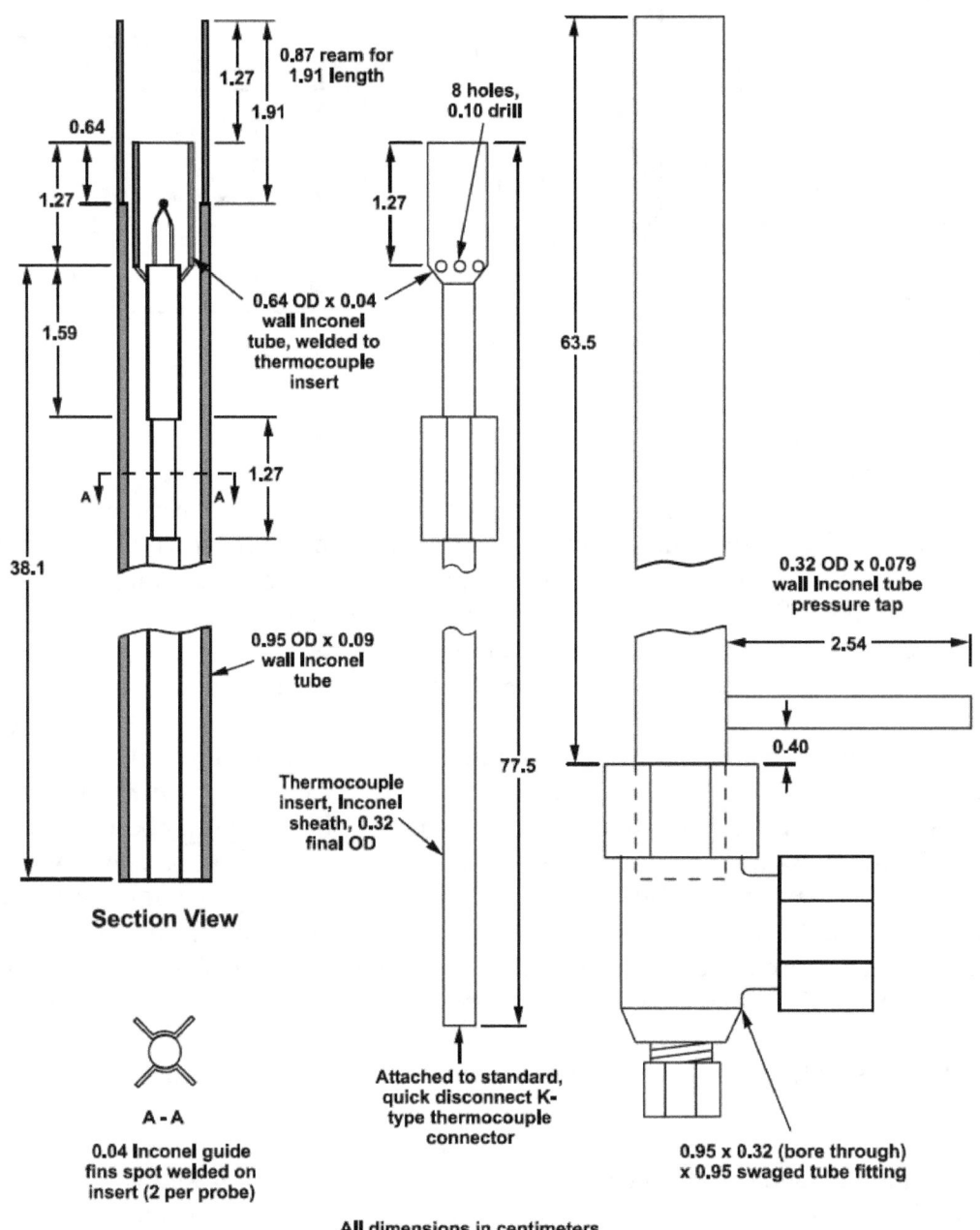

Figure 11. Detailed drawing of aspirated thermocouple using NACA design [34].

Each aspirated thermocouple probe was attached to the data acquisition system using K-type thermocouple wire and connectors. During each experiment, the flow meters and measured temperatures were monitored. These checks were performed in order to determine if any probe system became clogged so it could be unclogged with high pressure air. The difference in temperature signal between an inoperative probe and a properly flowing probe was obvious. A functioning aspirated thermocouple showed higher frequency temperature fluctuations due to the transient thermal environment and effective convective heat transfer while a non-functioning probe would not show rapidly fluctuating temperatures since the large mass of hot metal of the probe radiating to the bead and lack of convection would dampen any short fluctuations. A probe typically required about 1 min when activated to overcome accumulated heat and reach the true gas temperature.

To evaluate measurement uncertainty and instrument time response, the present study performed a series of detailed flow and heat transfer calculations, focusing on double-shielded aspirated thermocouples and bare-bead thermocouples. A detailed description of the calculations and results can be found in Appendix A.

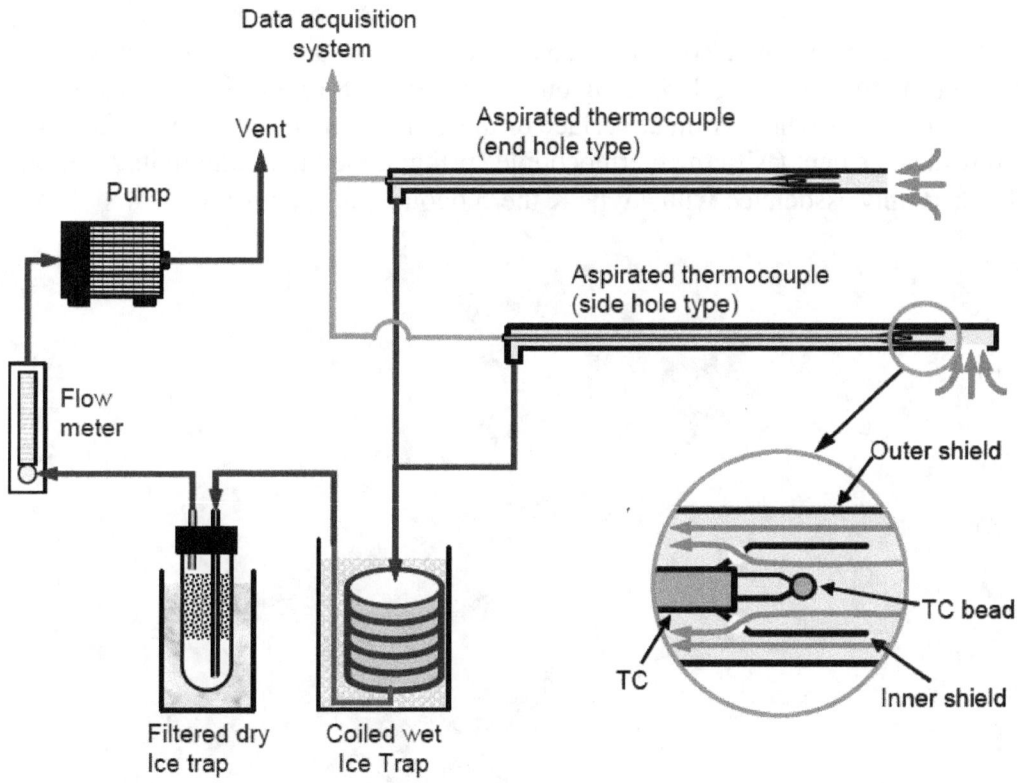

Figure 12. Schematic drawing of aspirated thermocouple measurement hardware.

Bare Bead Thermocouples
Temperature measurements by thermocouples located on compartment surfaces are required to characterize the thermal environment created by the fire, as well as to provide information for the aspirated thermocouple uncertainty analysis. Surface temperature measurements on the external surfaces of the compartment are needed to check the overall enthalpy balance associated with the

fire, which is important for validating predictive compartment fire models. Bare bead thermocouples were also used in the doorway adjacent to the bi-directional probes and aspirated thermocouples.

The bare bead thermocouples were created by removing 1 cm to 2 cm of silica ceramic yarn insulation from chromel and alumel (type K) lead wires and spot welding them together. The thermocouple lead wire diameter was 0.51 mm (24 gauge). The mean bead diameter was approximately two times the wire diameter.

Bare bead thermocouples were placed on the inside surface of the compartment at three locations: two on the floor and one on the ceiling. Table 2 in Section 2.4 lists the exact locations of the thermocouples, which were positioned on the floor, adjacent to the total heat flux gauges, and on the ceiling, almost directly above the front measurement station (which included the gravimetric soot and aspirated thermocouple probes). Figure 13 shows a 30 gauge Type K thermocouple on the compartment floor surface, which was held in place by spring loading to maintain its position near a 6 mm diameter total heat flux gauge. The screw/washer assembly about the thermocouple wire ensured that the thermocouple would not move.

Bare bead thermocouples were also positioned on the external surface of the compartment at one to five locations on the rear wall, depending on the experiment. Figure 14 shows a type K thermocouple attached to the rear outer surface of the compartment and held in place by a washer/screw arrangement (with the thermocouple spring loaded to maintain its position). The expanded uncertainty associated with a type K thermocouple is approximately 4.4 °C [35].

Figure 13. Type K thermocouple on compartment floor surface near a 6 mm diameter total heat flux gauge.

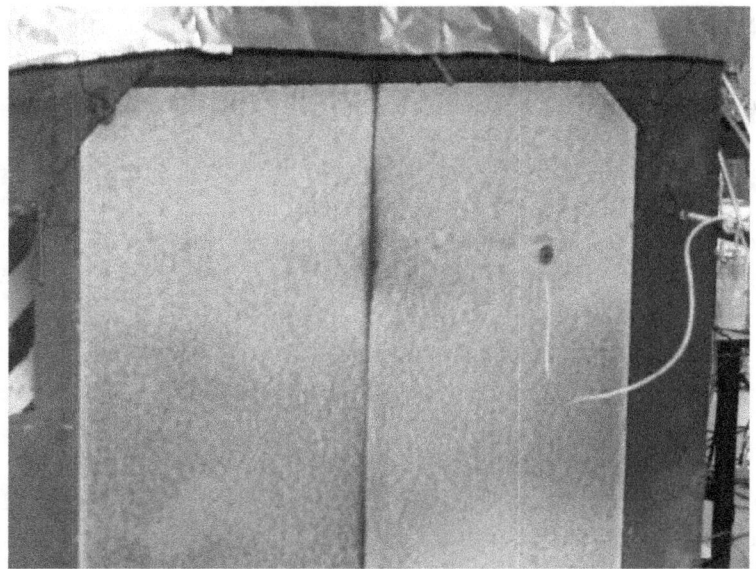

Figure 14. Type K thermocouple attached to the outer surface of the compartment.

2.5.6 Pressures and Velocities

Dynamic pressure was measured at 9 locations in the doorway of the enclosure in order to determine velocities in the doorway. The coordinates of the locations are contained in Table 3 and shown in Fig. 4. The differential pressure transducers (0 V to 10 V output) had a maximum range of 133 Pa. The particular model information is contained in Appendix D. Each pressure transducer was mounted on a board, and the board was attached to one of the support legs of the exhaust hood. Insulating board was also used to shield the transducers from the thermal insult of the fire. Each transducer produced a voltage, $V_{bdp,}$ related to the exposed differential pressure by the following equation: $\Delta P_{bdp} = 13.332(V_{bdp} - V_{bdp,zero})$, where the pressure difference is in pascals and the voltages are measured in volts. The zero voltage, $V_{bdp,zero,}$ condition is created when the positive and negative ports of the transducer are connected so there can be no pressure difference between them.

The transducers were connected to 1.3 cm diameter bi-directional probes [36] with 6.4 mm diameter copper tubing. Probe leads were routed close to each other so each lead was exposed to the same levels of heating. This installation care minimized differential heating and any resulting non-flow induced pressure differences between the leads.

Bi-directional probes enable the measurement of dynamic pressure which is the difference between the total pressure on the face where flow impinges and the static pressure on the downstream face of the probe. Using Bernoulli's principle and including a calibration factor, velocity, v, can be obtained from the dynamic pressure and a local gas temperature through the following relation: $v = C\sqrt{\Delta P_{bdp} T_{bdp}}$ where ΔP_{bdp} is the measured pressure across the bidirectional probe, T_{bdp} is the temperature (in K) of the gases flowing past the probe and C is defined as: $C = \dfrac{1}{C_{bdp}} \sqrt{\dfrac{2R}{P_{ref} MW_{gas}}}$.

The calibration coefficient, C_{bdp}, for a bi-directional probe is equal to 1.08 ± 0.05 [36] when the local Reynolds number (defined by the probe diameter) is greater than 1000. R is the ideal gas constant and MW_{gas} is the molecular mass of the gas.

To generate the velocity from the differential pressure, the temperature near the bi-directional probe is required. Because aspirated thermocouples can intrude on the pressure measurement bare-bead thermocouples were used to measure temperature. See Appendix A for a discussion of the errors associated with this approach. The measured doorway velocities were in the range of -7 m/s (flow out of the enclosure) to +1.5 m/s (into the enclosure). The combined expanded (k = 2) uncertainty in the velocity measurement varied from $\pm$ 0.5 m/s to $\pm$ 2.3 m/s. The largest component of uncertainty in this measurement was the variation in the pressure signal.

2.5.7 Gas Chromatography

A gas chromatograph was used at discrete times during the RSE tests to identify and quantify the major hydrocarbon species for each fuel and fire size at the front gas sampling location. The majority of the stable intermediate species were identified and quantified with a Hewlett-Packard 5890 gas chromatograph (GC) with flame ionization detector (FID). For chromatographic separation, a Restek Rt-QPLOT column (30 meter, 0.32 mm ID) was installed in the HP5890. Identification of the unknown species was accomplished by retention time matching. The quantification of the identified compounds was accomplished using the FID. The settings for the HP5890 are specified in Table 4.

In order to identify and quantify the hydrocarbons, gas phase calibration standards were obtained from Scott Specialty Gas. These standards had a reported uncertainty of $\pm$10 % of desired volume fractions. Several other sources of gas standards were available for only retention time matching of unknown compounds. Table 5 documents the retention times for the specific identified compounds.

To quantify the identified compounds, either the calibration curve for the specific identified molecule or a calibration curve of a similar molecule was employed. If the molecular mass of the identified molecule and calibration curve molecule were different, then a correction factor based on the calibration molecule's carbon number divided by the identified molecule's carbon number was used. Employing this technique with two known compounds, the relative uncertainty associated with this correction was approximately 1 % for one carbon atom difference, and approximately 3 % for two carbon atom difference. As a result, all correction factors for unknown compounds were generally limited to a 1 or 2 carbon number difference.

Table 4. Gas chromatograph settings and method parameters.

GC Parameters	Setting
Inlet Parameters	
Inlet Mode	Constant Pressure Rate
Inlet Pressure	124 kPa
Inlet Temperature	250 °C
Column Flow	2.3 mL/min
Split Ratio	10:1
Septum Purge	1.0 mL/min
Oven Temperature Parameters	
Initial Temperature	50 °C
Initial Time	1 min
Ramp #1 Rate	15 °C/min
Ramp #1 Temperature	80 °C
Ramp #1 Hold time	1 min
Ramp #2 Rate	20 °C/min
Ramp #2 Temperature	240 °C
Ramp #2 Hold time	5 min
Auxiliary Parameters	
Valve Oven Temperature	125 °C
FID Temperature	250 °C
Valve Timing Parameters	
0.10 min – Valve #A	ON (Sample Injection)
1.10 min – Valve #A	OFF

Table 5. List of retention times for identified compounds.

Species	Elution Time (min)	Species	Elution Time (min)
methane	2.41	n-butane	9.27
ethene	3.60	1-butyne	9.34
ethyne	3.62	2-butyne	10.22
ethane	4.07	1-pentene	11.00
propene	6.58	n-pentane	11.17
propane	6.80	1-hexene	12.79
propyne	7.12	n-hexane	12.93
1-butane	9.05	benzene	13.74
1,3-butadiene	9.18		

Analysis of variance (ANOVA) and regression analysis were employed in an effort to determine the errors associated with the quantification of the intermediate combustion species. The result of this analysis was the uncertainty of a single value, S_y, calculated from the calibration curve. The equations utilized for the analysis are show below:

$$S_y = \frac{S_{regression}}{\beta}\sqrt{\frac{1}{m} + \frac{1}{N} + \frac{(y_i - \bar{y})^2}{\beta^2 S_{xx}}} \tag{1}$$

where,

$$S_{regression} = \sqrt{\frac{\left(S_{yy}\right) - \left(\frac{(S_{xy})^2}{(S_{xx})}\right)}{N-2}} \tag{2}$$

$$S_{yy} = \sum y^2 + \frac{\left(\sum y\right)^2}{N} \tag{3}$$

$$S_{xy} = \sum xy + \frac{\left(\sum x \sum y\right)}{N} \tag{4}$$

$$S_{xx} = \sum x^2 + \frac{\left(\sum x\right)^2}{N} \tag{5}$$

m = number of measurements of the unknown sample

N = number of calibration curve points (typically, 3)

y = FID area count of calibration species

x = volume fractions of calibration species

y_i = FID area count of unknown species

β = slope of linear least squares fit to calibration points

The results of this analysis are included in all of the quantification graphs for each of the fuels examined in this program. However, it is important to note that this analysis does not include other sources of errors, such as sample extraction errors, possible decay of the samples prior to analysis, or errors resulting from carbon number correction factors.

2.5.8 Optical Extinction Soot Measurements

Two optical configurations were used to measure soot using laser transmission. They are shown schematically in Figure 15 and Figure 16, and the optical components used in both configurations are listed in Table 6. Results are available for Tests 6, 13, and 14 only, mainly because the measurement proved to be very challenging. The method and instrumentation used during Test 6 differed from that of Tests 13 and 14, as the experimental technique was being optimized during the course of this study. Table 7 lists the fuel type used during the tests. The apparatus used in Test #6 (shown in Fig. 15) consisted of a laser light source, a chopper, a beam splitter, purged tubes, optical lenses and two detectors. A chopper and lock-in amplifier were used to minimize the background signal and amplify the original signal. The reference signal of the original laser source was monitored by splitting the laser beam.

Figure 16 shows the apparatus used during tests #13 and tests #14. The laser light source was a 9.7 mW 657 nm diode laser (continuous). A spherical lens with a 2.5 cm diameter and 40 cm focal length reduced the beam diameter. The optical path-length was defined by two thin (1 mm wall thickness) 2.5 cm diameter stainless steel tubes that were aligned and separated by a gap of distance L_D. Each of these tubes was connected to commercially available "stackable lens tubes" that were used to align the laser and detector and to shield the detector from stray light. Low flowing nitrogen gas was used to prevent combustion products, including soot, from entering the tubes. A glass window was attached on the end of the stackable lens tubes to prevent backward flow of the nitrogen purge gas. The detector was a silicon photodiode sensitive from 350 nm to 1100 nm. On the detector side of the optical assembly, a spherical lens (2.5 cm diameter and 1 m focal length) was positioned in the middle of stackable lens tubes to focus the laser light on the detector and prevent backward flow of the nitrogen purge gas. Several optical components were placed in front of the detector. A band pass filter (± 10 nm window centered about 650 nm) acted to attenuate radiation other than that from the laser. A diffuser lens was used to expand the laser beam, a standard practice that reduces beam steering effects. To prevent detector saturation, a neutral density filter was used to attenuate the beam.

The position of the optical apparatus used in tests #13 and #14 was different from that used in Test 6. The sampling position was located 1 cm below the doorway soffit (80 cm above the floor) and 1.4 cm from the plane defining the doorway as shown in Fig. 16. The nominal pathlength (L_D) (the distance between the stainless steel purging tubes) in the optical soot measurement was 11.5 cm. The nitrogen purge flow rate was adjusted to keep combustion products and smoke out of the purge tubes. The purge was measured as 900 ($\pm$30) cm^3/min, equivalent to an average speed of 4 cm/s through each of the tubes. The purge flow was estimated to reduce the actual path-length of the optical measurement (L) by approximately 1.0 cm. This value was observed to vary somewhat with fire size, such that the estimated expanded uncertainty of L_D was about 1 cm, corresponding to a relative expanded uncertainty of 9 %.

The determination of soot density (m_s) from an optical extinction measurement is based on Bouguer's law, in which the light extinction coefficient (K) is defined in terms of the attenuated intensity (I), and the reference intensity (I_o) of monochromatic light passing through a homogeneous smoke path of distance (L):

$$K = -\frac{\ln(I_o / I)}{L} \tag{6}$$

$$m_s = \frac{K}{\sigma_S} \tag{7}$$

where σ_s is the mass specific extinction coefficient. The recommended value of σ_s for flame generated smoke in over-ventilated fires is 8.7 m^2/g $\pm$ 5.4 % (standard relative uncertainty). The expanded uncertainty of the $\ln(I_o/I)$ term during a period of 6 min before the fire test was on the order of 0.2 %. The laser drift, which was about 5 %, dominated the uncertainty in the baseline.

The expanded combined uncertainty of the soot volume fraction is computed as the square root of the sum of the individual standard uncertainties [$u(x_i)$] associated with each of the terms that influence the soot mass measurement. The combined uncertainty of the optical soot density measurement is calculated as follows:

$$U_{m_s} = 2 \sqrt{\left(\frac{\partial m_s}{\partial \ln(I_o/I)}\right)^2 u\big(\ln(I_o/I)\big)^2 + \left(\frac{\partial m_s}{\partial L}\right)^2 u(L)^2 + \left(\frac{\partial m_s}{\partial \sigma_s}\right)^2 u(\sigma_s)^2} \qquad (8)$$

An estimate of the combined relative expanded uncertainty was about 20 %, for both experimental configurations.

Table 6. Components used in the optical extinction measurements.

Component	Test 6	Tests 13 and 14
Spherical lens before detector	Present	Present
Spherical lens after laser	Not Present	Present
Beam splitter	Present	Not Present
Neutral density filter	Not Present	Present
Reference detector	Present	Not Present
Aperture in the detector part	Not Present	Present
Lock-in amplifier	Present	Not Present
Probe separation distance (L_D)	$\gg 24$ cm (≈ 6 m)	11.5 cm $\pm$ 0.1 cm
Probe location	5 cm below door soffit; Across entire doorway; 1.4 cm beyond compartment	1 cm below door soffit; about center of doorway; 1.4 cm beyond compartment

Table 7. Tests in which optical light extinction measurements were conducted.

Test [1]	Fuel type	Burner [2]	Fire size, Q_{max} (kW)	Path-length, L (cm)
6	Natural Gas	B (25 cm square)	75 to 400	24.0 $\pm$ 1.5
13	Polystyrene	D (20 cm round)	18	10.5 $\pm$ 1.0
14	Polystyrene	E (40 cm round)	80	10.5 $\pm$ 1.0
1. measurements not made in other tests, because the fire adversely affected the detection system. 2. see Fig. 2.				

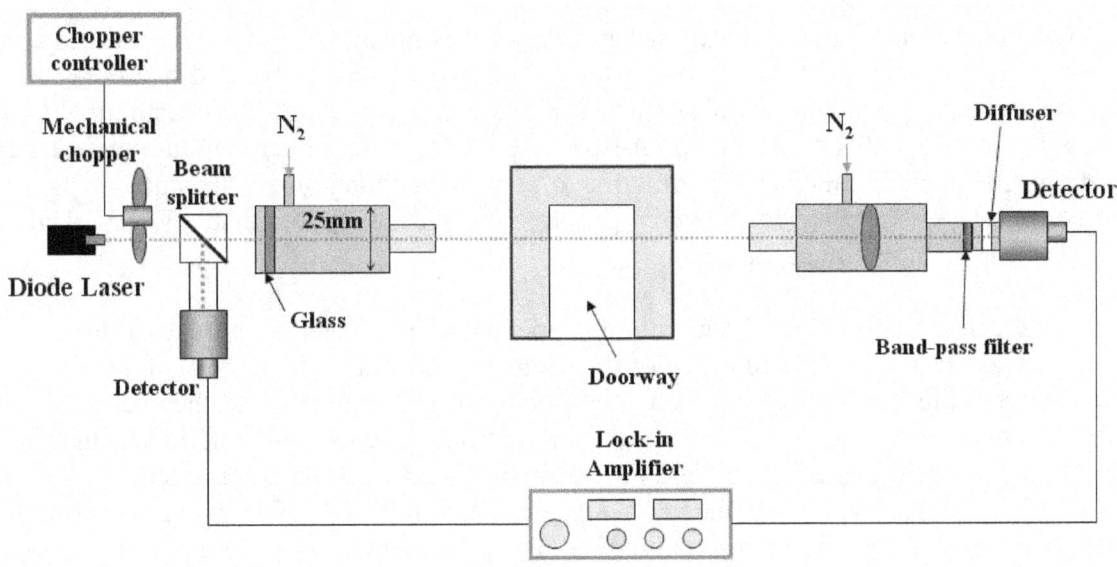

Figure 15. Schematic of the optical extinction measurement apparatus used during test #6.

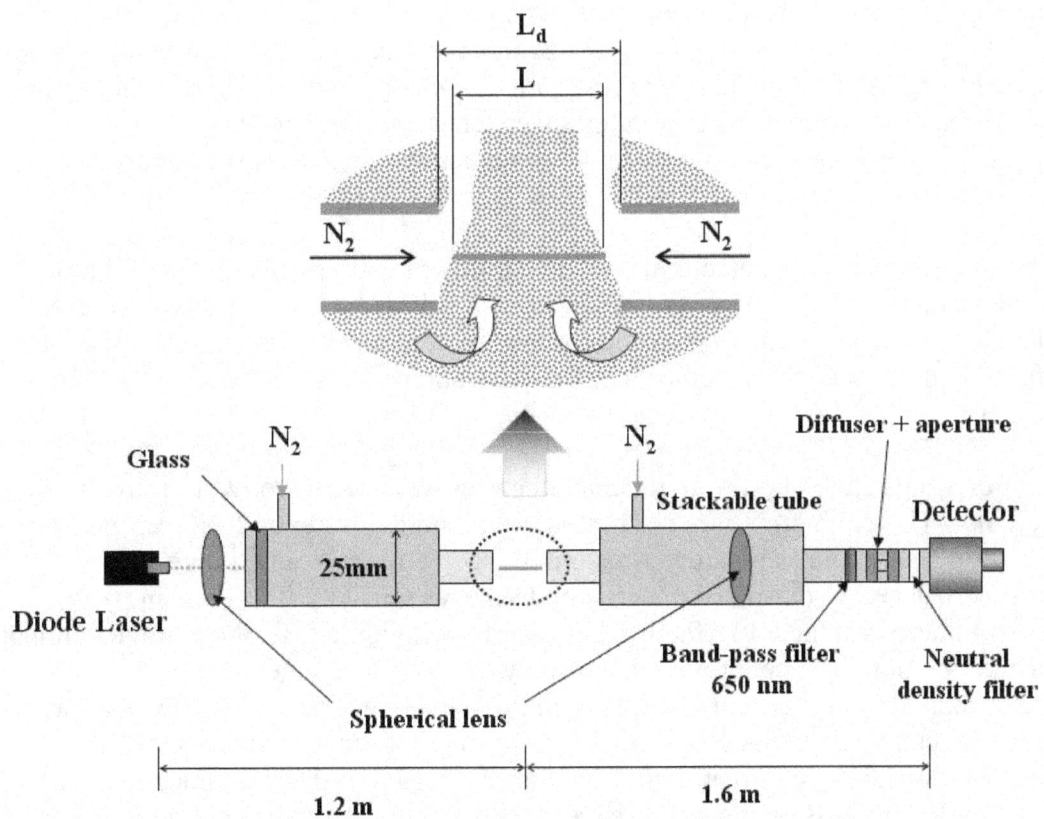

Figure 16. Schematic of the optical extinction measurement apparatus used during test #13 and test #14 including a blow-up of the circled optical sampling region.

2.5.9 Heat Fluxes

Total heat flux was measured at two locations during each experiment. The heat flux gauges were 6.4 mm diameter Schmidt-Boelter type, water cooled gauges with embedded type-K thermocouples. The particular model information is contained in Appendix D. The nominal range for the gauges was 150 kW/m^2. Schmidt-Boelter gauges measure a temperature difference across a thin insulating material using a thermopile to generate a voltage from the small temperature difference. These gauges typically generate voltages much less than 100 mV even for heat fluxes near their maximum range.

Each gauge was inserted in the floor flush with the upper surface and facing vertically upward. The rear gauge was located on the centerline of the enclosure and approximately ¾ of the way toward the rear from the front. The front gauge was located on the centerline of the enclosure and approximately ¼ of the way toward the rear from the front. The exact location coordinates for the gauges are listed in Table 2. The condition of the installed gauges was checked periodically. If significant soot accumulated on a gauge, it was brushed off. If a gauge was no longer flush with the surface of the floor, a note was made, but there was no attempt to move the gauge since the gauges were very difficult to access and attempting to do so could have impacted the integrity of the floor.

Heat fluxes as high as 250 kW/m^2 were observed. These heat fluxes are beyond the stated range of the gauges. According to the manufacturer, the calibrations remain linear and valid beyond the stated range as long as the materials do not degrade and change the sensitivity of the gauge. After the first six experiments the heat flux gauges were checked for changes to their calibrations. Each gauge's responsivity was found to remain within 3 % of the factory calibration.

The main sources of uncertainty related to the total heat flux measurements are: the calibration, soot and dust deposition, and shifting of the gauge surface below the floor. These sources will be described and the total uncertainty estimated for the reported measurements. A model of uncertainty for heat flux gauge measurements in fire environments can be found in the study by Bryant et al. [37].

The total heat flux gauge calibration from the manufacturer was used to convert millivolt readings to kW/m^2. This calibration was performed using cooling water at 23 °C ± 3 °C. The cooling water in the Large Fire Laboratory was found to be within the same range. The manufacturer reported a ±3 % expanded uncertainty in the responsivity (the slope in kW/m^2/mV). Calibrations at the NIST facility have varied within the 3 % range of the nominal manufacturer's calibration. A recent round-robin study of heat flux gauge calibration consistency [38] sent the same heat flux gauges to multiple laboratories around the world and found that while several calibrations fell within the 3 % range, if some outlier data were included, then the uncertainty rose to around 8 %. For this current project, an uncertainty of ±6 % for gauge calibration was chosen as fairly conservative since the NIST calibration was within the 3 % range in the round-robin study.

While the cooling water was supplied at approximately 23 °C, the fire heated the water such that the gauge temperature typically rose to between 40 °C and 60 °C, and less frequently to 100 °C.

For the fires where the water temperatures increased to between 40 °C and 100 °C, the heat fluxes were on the order of 100 kW/m^2 to 300 kW/m^2 which represent blackbody temperatures in the 950 °C to 1300 °C range. The most extreme combination (affecting uncertainty) of cooling water and environment temperature would be a 75 °C increase in cooling water in a 950 °C environment. This combination would only have about a 0.5 % effect on the measured heat flux. The effect was determined by calculating the ratio of the T^4 difference between 950 °C and the 25 °C cooling water with 950 °C and the 100 °C cooling water. This is a simplified comparison which assumes everything else is equal, but generates an approximation of the magnitude of the cooling water effect under specified conditions.

Heat flux uncertainty due to soot and dust deposition is difficult to quantify. For many tests, such as those burning methanol, ethanol, and natural gas, there was little to no contact with soot or combustion products. Also, even for the sootier fuels at low heat release rates, the lower layer remained as air with little opportunity for soot-laden gases to contact the gauges. For those experiments with sooty fuels and underventilated conditions (>200 kW HRR), combustion products including soot sometimes impinged on the floor. For these periods of time, it was estimated that the soot coating on the gauge would add an additional uncertainty of ±10 % due to variations in surface emissivity, and soot agglomerates shadowing the surface of the gauge.

The physical shifting of the gauge surface below the floor could have impact on a heat flux measurement if the solid angle viewable by the gauge was significantly diminished. Since the gauge is not sensitive either in calibration or application to radiation at angles close to the plane of the gauge surface due to reflection, and the radiation approaching from the lowest angles is generally from the coolest regions of the enclosure, the gauge would have to be below the surface of the floor by a few millimeters or more for there to be a significant impact on its measurement. Neither gauge was ever observed to be shifted by that amount in the course of testing.

2.6 Data Acquisition

Data acquisition (DAQ) for this series of experiments was divided into two systems. One DAQ system was dedicated to fuel flows, oxygen depletion calorimetry, and the constituent measurements required to calculate heat release rate using that method. The other DAQ system was used to record signals from all other measurements. Each DAQ system used National Instruments hardware and was controlled with LabVIEW software. The calorimetry DAQ system has been previously described in detail [29].

For this series of experiments, the channel list contained in Appendix C was used to program the DAQ system. The types of measurements included: gas analyzers, dew point readers, heat flux gauges, pressure transducers, and thermocouples. These measurements were recorded on the DAQ hardware as voltages with 200 samples recorded every second. Each second, the average value for each channel was then converted to meaningful physical units. Two event marking channels were used to note the time of important events such as ignition, fuel flow change, or extinguishment. These event marker channels, which are in both DAQ programs, were especially useful in synchronization of the two data sets.

The DAQ system for measurements not related to heat release rate, called MIDAS (Modular In-situ Data Acquisition System), had a structure and hardware components that differ from the HRR/fuel DAQ system. The MIDAS system utilized a fiberoptic extension of the computer's PCI bus. A series of fiberoptic cables connected the main computer in the control room to three experiment stations in the main bay of the laboratory. Each station has its own DAQ card, multiplexing hardware, and terminal blocks for voltage or thermocouple inputs. The East MIDAS Station was the station used for this series of experiments.

There were four comma-delimited spreadsheet files produced for each experiment. One file (with the -raw suffix) contained all of the raw voltages and temperatures recorded. A second file (-adj suffix) contained values with converted units calculated from the raw voltages. A third file (-ZS suffix) contained calibration data for each instrument that is calibrated at the beginning of an experiment. Finally, a fourth file (-sd) contained standard deviations for selected instruments based on the 200 raw voltages averaged each second.

The data acquisition hardware had 16 bit precision, with stated accuracies of the data acquisition board and multiplexing module equal to 0.014 % and 0.015 % of the reading. These uncertainties were orders of magnitude lower than those from other sources in all of the measurements reported here.

2.7 General Data Corrections
A Matlab script file was created for post-processing all data files generated during the test series. This program was used to make corrections to the data, generate plots, and save results to ASCII text files for archival purposes. The program was also used to compute time averaged values and uncertainties for examining trends in the data. An input file was used to allow batch processing of the raw data files. The input file contained the parameters needed for the heat release calculation (this file was also read by the DAQ program during the data collection process). Additional parameters were added to the end of the standard HRR input file to account for the gravimetric soot measurements and to record the time windows when channels had known missing or corrupted data.

The first step in data reduction was to inspect the data files and lab notebooks for erroneous data resulting from open channels, loss of sample flow, or some other instrument or data acquisition malfunction. Because data were collected on two separate computers, the series were synchronized to a common reference time. The ignition time was marked using a virtual event channel on each computer and defined as time zero for the reduced data. The gas analyzer measurements from inside the RSE and exhaust hood measurements were shifted in time to account for the sample flow transfer (delay) time. There was no adjustment for instrument response time in the data reduction. . The t_{10-90} response time of the instruments used in this study varied from 1 s to 12 s.

Corrections to the heat release rate measurements were applied to account for the exhaust flow calibration factor and drift in the oxygen analyzer. The exhaust flow rate data was smoothed over a 10 s window to reduce noise due to turbulent flow in the duct. This smoothing was of the same order as the response time of the exhaust gas sample oxygen measurement.

Since the gases sampled from the RSE were dried before entering the detectors, an estimate of the water removed must be made in order to correct the measurements to the *in situ* wet volume fraction. The general combustion reaction assuming all the fuel is reacted and that the soot can be represented as pure carbon is:

$$C_x H_y O_z + a O_2 \rightarrow b CO_2 + c CO + d CH_4 + e C + f H_2 O \tag{9}$$

The molecular yield of water can be related to the combustion product yields using the known hydrogen/carbon (y/x) ratio of the fuel:

$$f = \frac{y}{2x}(b + c + d + e) - 2d \tag{10}$$

If the yield of soot is small compared to the other products, the water volume fraction, X_{H2O}, can be estimated from Eq.11.

$$X_{H_2O} = \frac{y}{2x}\left(X_{CO_2,wet} + X_{CO,wet}\right) \tag{11}$$

The relationships for wet CO_2 and CO are given by the following:

$$X_{CO,wet} = \frac{X_{CO,dry}}{1 + \dfrac{y}{2x}\left(X_{CO_2,dry} + X_{CO,dry}\right)} \tag{12}$$

$$X_{CO_2,wet} = \frac{X_{CO_2,dry}}{1 + \dfrac{y}{2x}\left(X_{CO_2,dry} + X_{CO,dry}\right)} \tag{13}$$

Other gas volume fraction measurements performed on a dry basis were corrected using the following relationship:

$$X_{spec,wet} = X_{spec,dry}\left(1 - X_{H_2O}\right) \tag{14}$$

The total hydrocarbons can contribute to the formation of water, however the gas composition measurements confirmed that when total hydrocarbons were present in significant quantities, they were in the form of unburned fuel (methane in the natural gas tests). Unburned fuel does not contribute to the formation of water. Therefore, the resulting relative error in the water volume fraction estimation due to neglecting hydrocarbons was always less than 3 %. The error in the water volume fraction estimate due to neglecting soot was as much as 10 % for the highly sooting fuels. However, since the soot measurements were sparse, we chose to report the results on a consistent basis. A more accurate estimate of water volume fraction could be made for the short time windows where soot was collected. Hydrogen gas was not quantified for these tests, but could also affect the estimation of water.

Unless otherwise noted, all uncertainty results reported here represent the combined expanded (coverage factor k = 2) uncertainty resulting from a propagation of uncertainty analysis. The uncertainty values are represented by error bars on the steady state average values presented in Sec. 3.

3 Results

3.1 Heat Release Rate

The heat release rate measurement was used to characterize the size of the fire and also to help determine (along with heat flux data) when the fire conditions had reached steady state. As the fire becomes underventilated burning can take place outside of the enclosure. The HRR measurement represents the total burning inside and outside of the enclosure. Table 8 shows a description of the measurement labels used in the table column headings and figure legends in this section. These labels are identical to the column headings in the reduced data files.

Table 8. Description of calorimetry measurement labels.

Measurement Label	Description
54 HRRcal	Heat Release Rate from Calorimeter, kW
55 HRRburner	Heat Release Rate from Burner (gas, pool or spray), kW
56 StackMFR	Exhaust hood mass flow rate, kg/s
57 Tstack	Exhaust hood temperature (near bi-directional probe), °C
58 O2stack	Exhaust O_2 volume fraction (dry)
59 CO2stack	Exhaust CO_2 volume fraction (dry)
60 COstack	Exhaust CO volume fraction (dry)
61 THCstack	Exhaust total hydrocarbons volume fraction (dry)
62 MSstack	Exhaust soot mass concentration (wet), mg/m^3

Figure 17 shows the heat release rate results for one of the natural gas experiments (test #3). For this test, the measured HRR and enthalpy input match closely indicating that the combustion efficiency was close to 1. The composition and heating values for the natural gas used during all of the tests is shown in Table 9. The heating value is defined at the standard conditions of 300 K and 101.3 kPa. The propane and nitrogen levels were unusually high on the first test day; however the variation in heating value was less than 1.5 % for all of the tests with natural gas. Figure 18 shows two photographs of test #3 (natural gas) looking into the doorway at the nominal fire sizes of 75 kW (left) and 400 kW (right). The intensity and transparency of the flames provided visual evidence of the lack of soot produced by the fire. The image of the 400 kW fire clearly shows flames exiting the doorway, indicating the fire is underventilated. The 75 kW fire in the reduced-scale enclosure would scale up to a 485 kW fire in the ISO 9705 room, and the 400 kW fire would correspond to a 2.6 MW fire in the ISO room (see Sec. 6).

The heat release results for the methanol spray fire (burner C) test #12 are shown in Fig. 19. The dashed line in this figure represents the ideal heat release rate of the fuel based on the delivered fuel flow rate. The liquid fuel height in the catch pan varied during the test, and once the fuel flow was stopped the burnout time was more than 20 min (see Fig. 19 from 3400 s to 4900 s). This accumulation effect explains why the heat release rate of the fire measured by calorimetry was significantly different than the set burner heat release rate. Although the HRR measurement had a larger uncertainty than the fuel flow rate measurement, in some cases it was a more reliable measurement of the actual fuel burning rate. Photographic images of three different methanol fires are shown in Fig. 20. The image on the left shows a 50 kW methanol pool fire burning outside of the enclosure in a 40 cm diameter pan prior to test #12. The middle and right images show methanol spray fires from test #12 at HRR's of 70 kW and 300 kW, respectively.

Table 9. Composition (volume fraction %) and heating values of natural gas used in RSE tests.

Test Name	Methane	Ethane	Propane	i-Butane	n-Butane	i-Pentane	n-Pentane	C6+	Nitrogen	CO2	MW_{ng} (g/mol)	HOC_{ng} (MJ/m^3)
RSE1NG	86.92	3.84	3.57	0.13	0.15	0.05	0.03	0.07	4.40	0.84	18.56	34.07
RSE2NG	93.45	3.79	0.72	0.12	0.15	0.04	0.03	0.07	0.79	0.84	17.32	33.78
RSE3NG	93.39	3.71	0.71	0.12	0.15	0.05	0.04	0.07	0.83	0.93	17.34	33.73
RSE6NG	93.95	3.07	0.70	0.12	0.15	0.05	0.04	0.08	0.83	1.02	17.28	33.54
RSE65NG	93.46	3.76	0.86	0.12	0.16	0.04	0.03	0.05	0.78	0.74	17.31	33.86

Figure 21 shows the heat release rate results for test #16 with polystyrene. Six kilograms of polystyrene pellets were ignited using a heptane spray fire which remained on for a period of 30 s. Figure 22 shows images of the fire at two different heat release rates. The left side of the figure shows the fire approximately 6.5 min after ignition when the HRR was 170 kW. Large amounts of black soot can be seen exiting the doorway. The right side of the figure shows the fire 8.5 min after ignition, when the fire size was 340 kW, approaching its peak HRR.

Figure 23 shows the heat release rate results for a toluene pool fire (test #10). The pool (burner B in Fig. 2) was half-filled with fuel when ignited at t = 0 s. As the thermal environment of the compartment changed, the fuel flow rate and the water flow rate to the liquid-cooled burner were continually adjusted to achieve near-steady burning. Figure 24 shows images of the toluene fires at heat release rates of 60 kW (left side) and 200 kW (right side).

The HRR results for the ethanol pool fire (test #11) are shown in Fig. 25. As with the methanol fire, the discrepancy between the measured HRR and the enthalpy input was primarily due to fuel accumulation in the pool. Images of test #11 with ethanol are shown in Fig. 26.

The HRR results for the heptane spray fire (test #15) are shown in Fig. 27. Unlike the tests with alcohol fuels, there was no evidence of fuel accumulation in the burner during test #15. Inspection of the HRR curve at approximately 3200 s after ignition shows that when the fuel delivery was reduced (from 500 kW to 90 kW), there was no lag in the HRR measurement. This suggests that incomplete combustion (rather than accumulation) is responsible for the significant differences in the measured and complete burning rates of the fuel. Images of the 160 kW and 370 kW heptane fires are shown in Fig. 28.

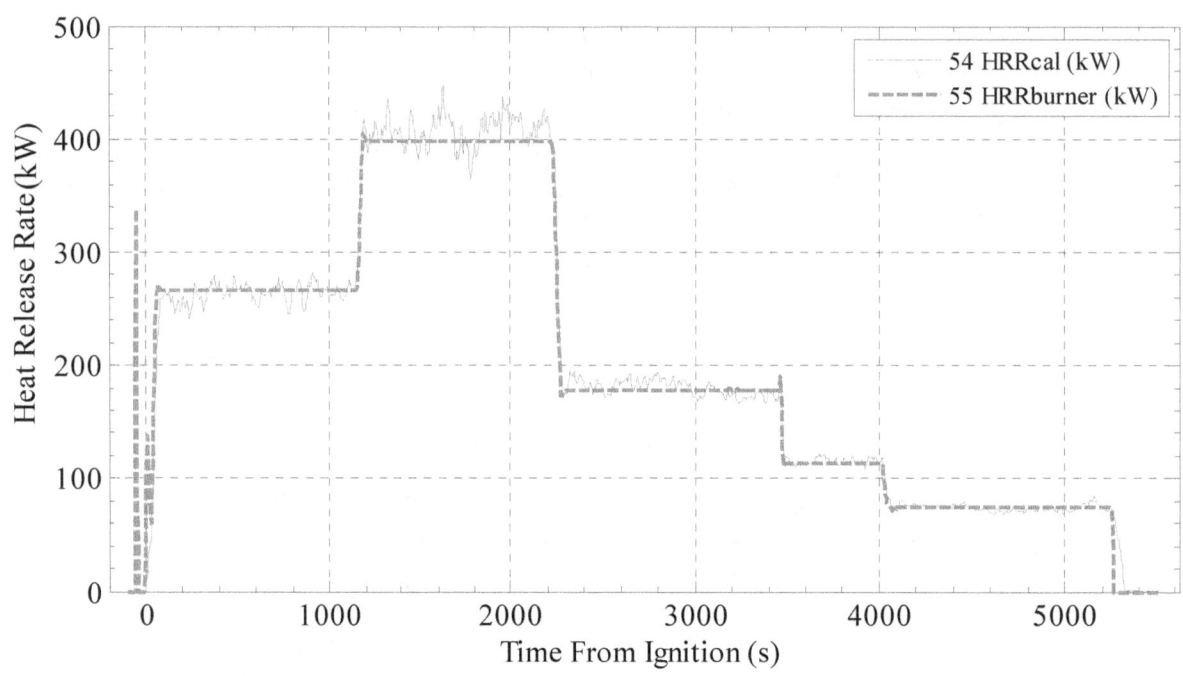

Figure 17. Heat release rate results for natural gas test #3 using Burner A.

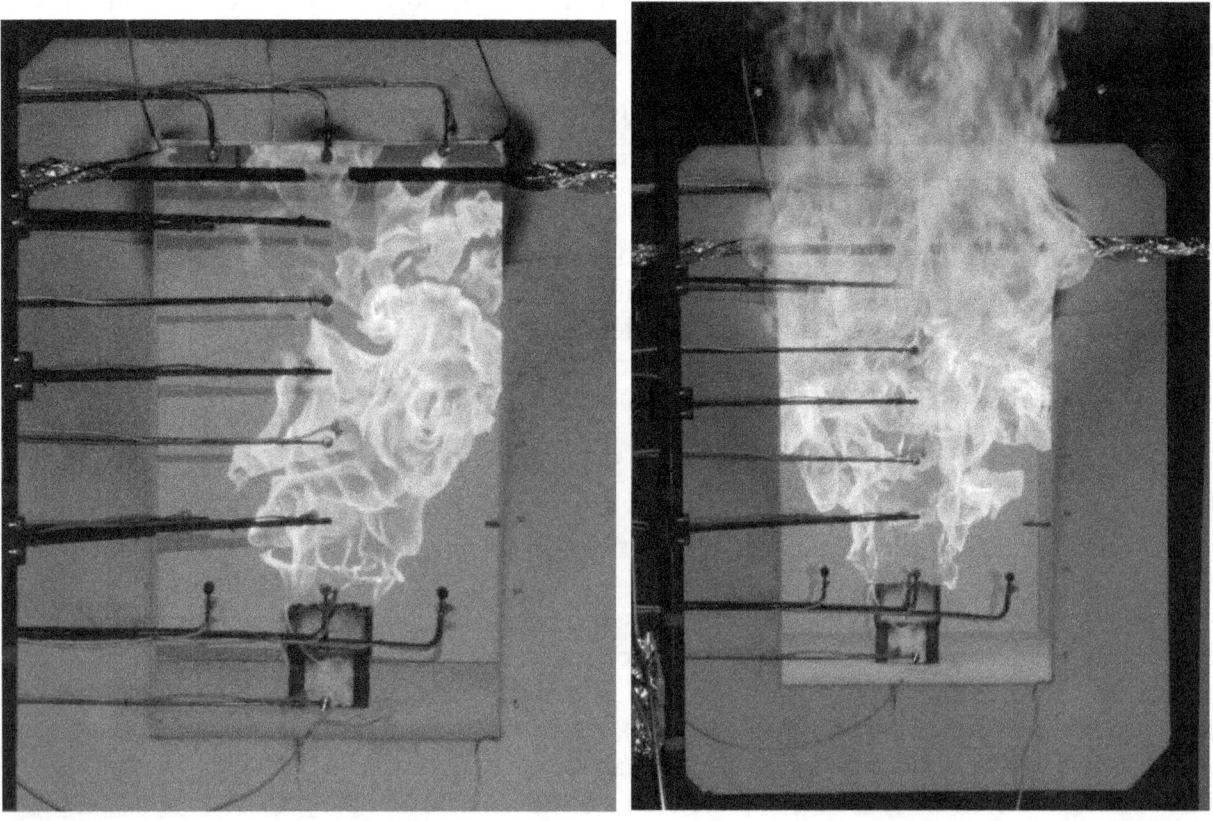

Figure 18. Photograph of test #3 natural gas, HRR = 75 kW (left), HRR = 400 kW (right).

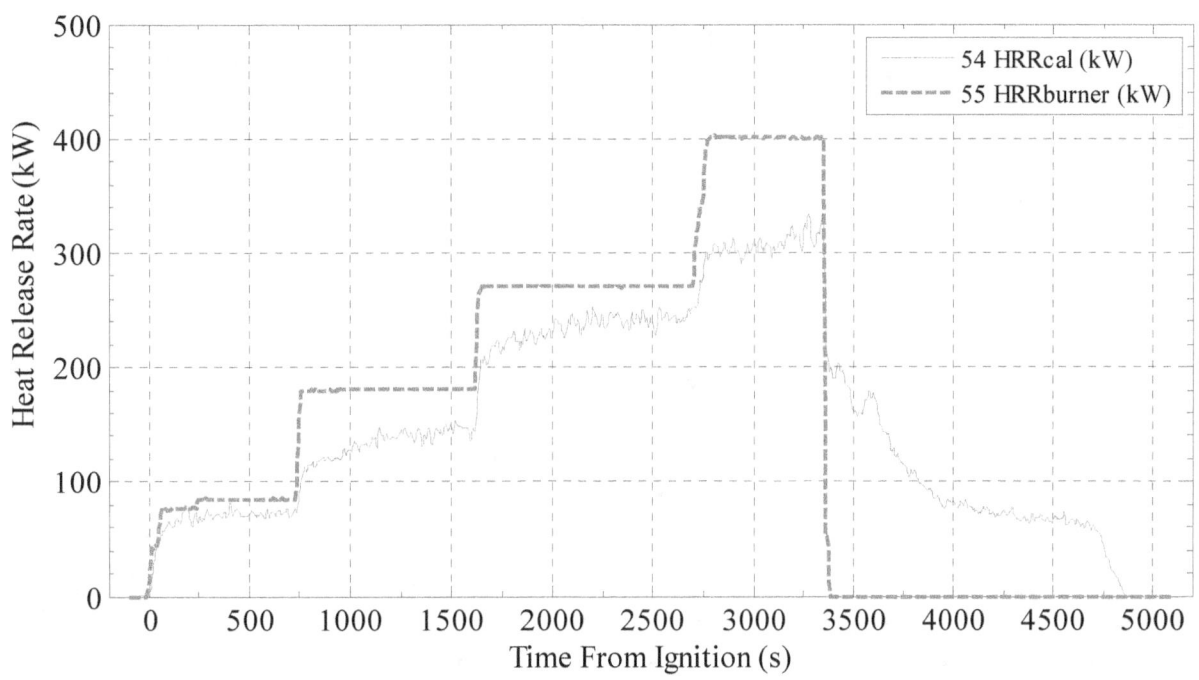

Figure 19. Heat release rate results for methanol test #12 using spray burner C.

Figure 20. Photographs of methanol test #12: Open burn, HRR = 50 kW (left), t = 645 s, HRR = 72 kW (middle), t = 2885 s, HRR = 305 kW (right).

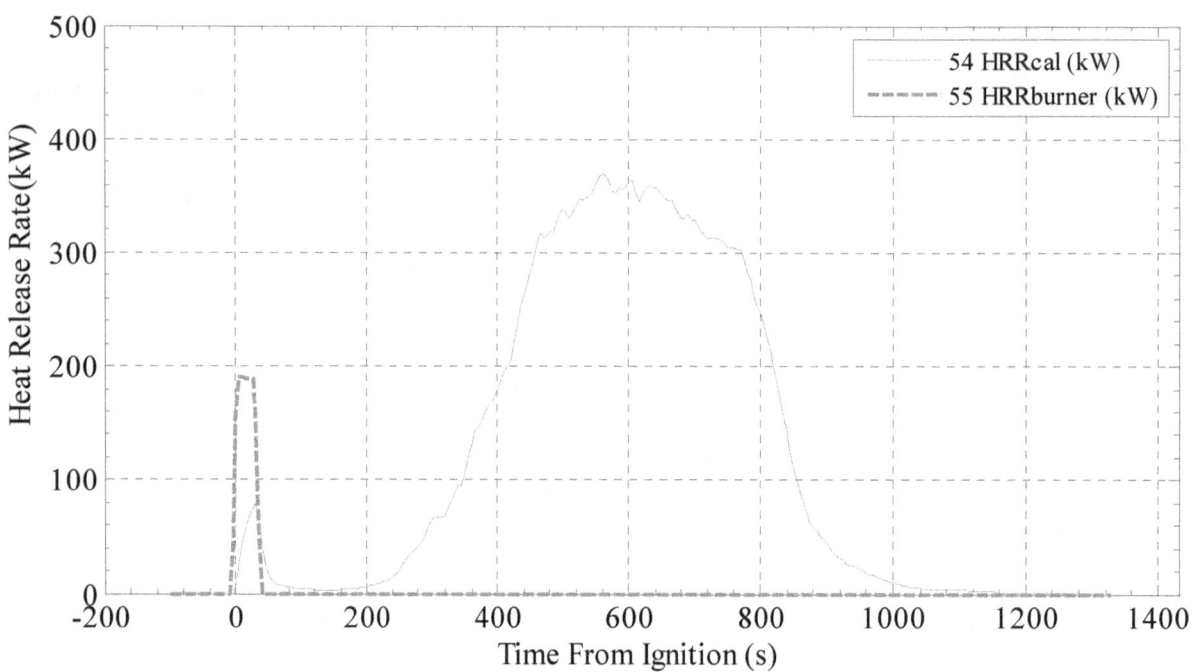

Figure 21. Heat release rate results for polystyrene test #16. A heptane spray was used to ignite 6 kg of polystyrene pellets.

Figure 22. Photographs of polystyrene test #16: t = 389 s, HRR = 170 kW (left), t = 515 s, HRR = 340 kW (right).

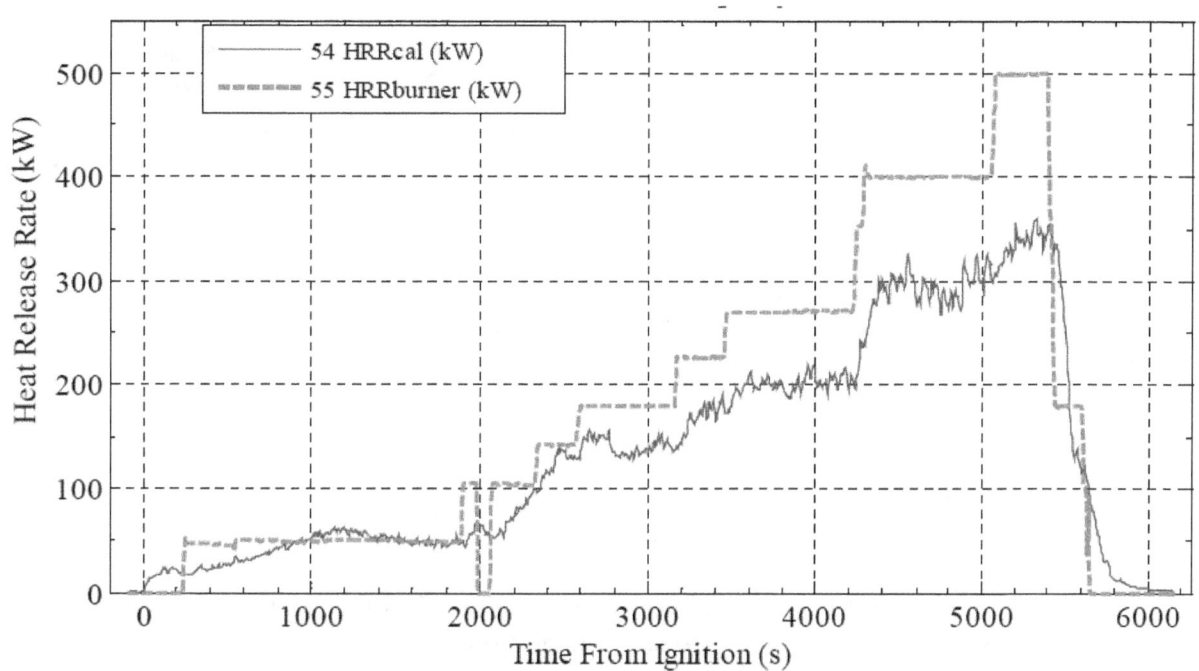

Figure 23. Heat release rate results for toluene pool fire (test #10).

Figure 24. Photographs of toluene pool fire (test #10): t = 1279 s, HRR = 60 kW (left),
t = 3843 s, HRR = 200 kW (right).

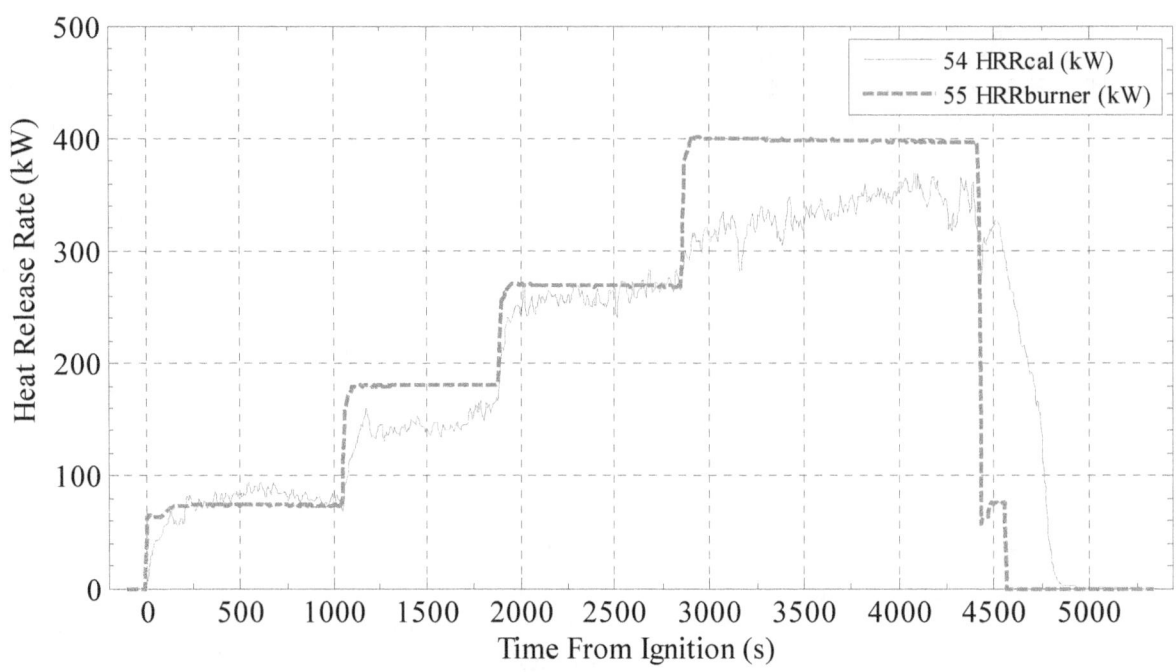

Figure 25. Heat release rate results for ethanol test #11.

Figure 26. Photographs of ethanol test #11: t = 1709 s, HRR = 145 kW (left),
t = 3328 s, HRR = 328 kW (right).

38

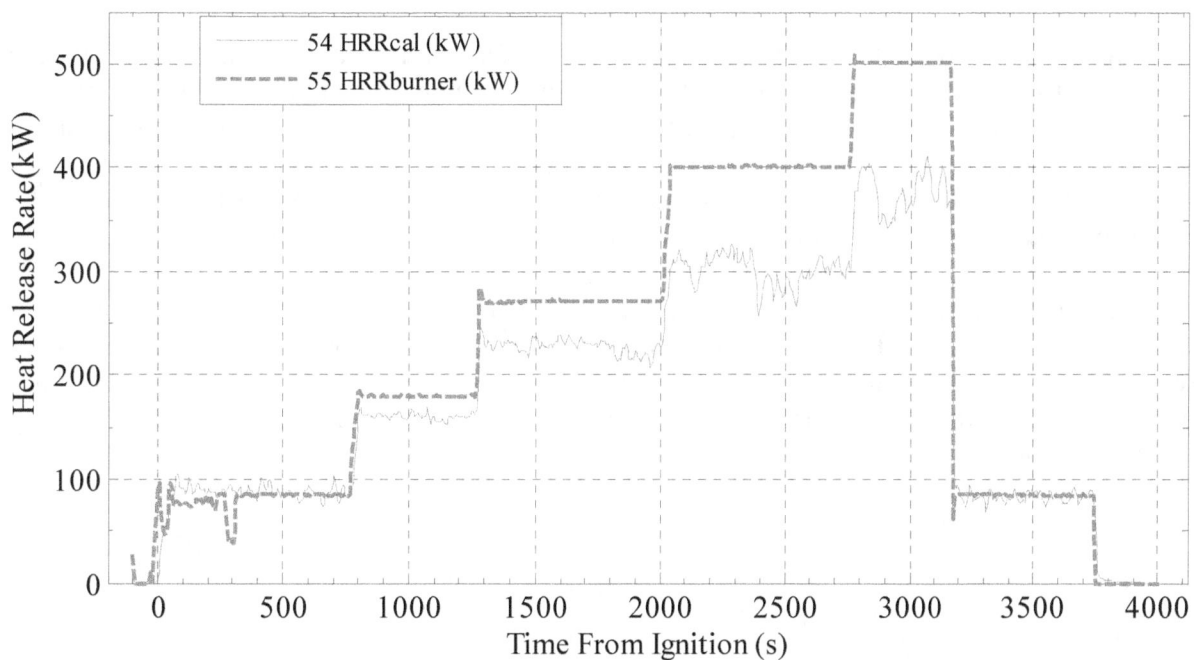

Figure 27. Heat release rate results for heptane test #15.

Figure 28. Photographs of heptane test #15: t = 1000 s, HRR = 160 kW (left),
t = 2870 s, HRR = 370 kW (right).

39

A summary of HRR results is shown in Fig. 29. The measured heat release rate of the fire using oxygen calorimetry is plotted as a function of the ideal heat release rate predicted by the set fuel delivery rate. The dashed line on this plot represents a combustion efficiency of 1 (complete combustion). As expected, the combustion efficiency of the cleaner burning fuels (natural gas, methanol and ethanol) was closer to 1.0 than the highly sooting fuels (toluene and heptane). For most of the fuels, the global combustion efficiency decreased as the fire became more underventilated. Further discussion of the combustion efficiency can be found in Sec. 5.5 of this report. A summary of the time averaged steady-state results of HRR and exhaust stack species measurements is given in Table 10. The averaging period for each row of data in this table is given in the column labeled "SS Window". A description of the remaining columns are given in Table 8. Total hydrocarbons in the exhaust stack were not measured in the exhaust stack during the first set of tests.

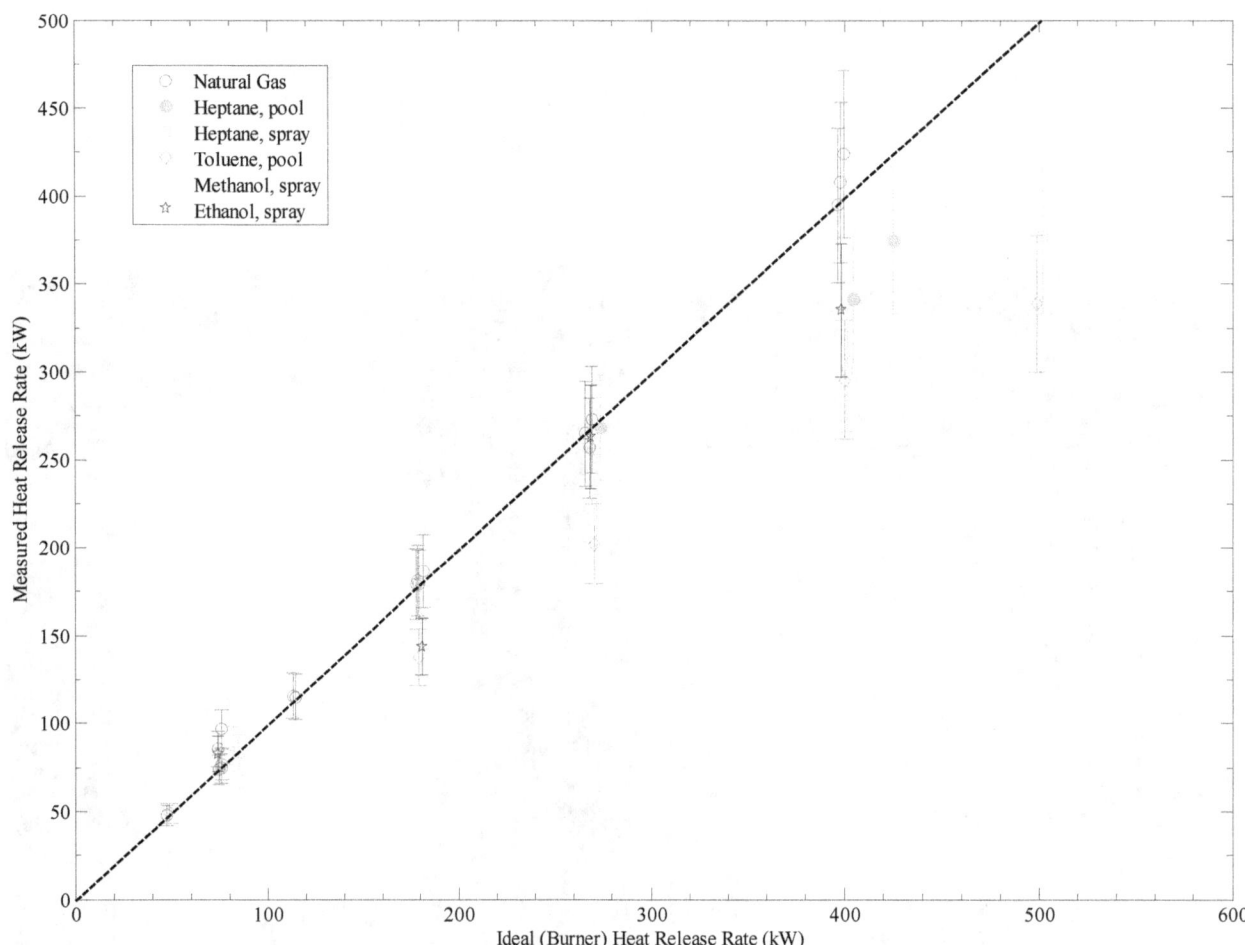

Figure 29. Steady state heat release rate results. Dashed line represents ideal or complete burning.

Table 10. Summary of time averaged steady-state results of HRR and exhaust stack species measurements.

Test #	Fuel	SS Window start (s)	stop (s)	HRRcal (kW) Mean	U	HRRburner (kW) Mean	U	O₂ stack (mol/mol) % Mean	U	CO₂ stack (mol/mol)*1e6 Mean	U	CO stack (mol/mol)*1e6 Mean	U	THC stack (mol/mol)*1e6 Mean	U	MS stack (mg/m³) Mean	U
1	Natural Gas	500	1924	74.6	9.1	75.7	2.4	20.75	0.21	1897	98	32	15				
		2220	3384	186.4	22.0	181.8	5.7	20.43	0.21	3780.0	194.4	59.2	26.5				
		3600	3944	77.0	10.0	75.9	1.7	20.74	0.21	1921.1	78.5	32.2	11.0				
2	Natural Gas	400	1594	256.9	30.3	268.1	4.6	20.14	0.20	5133.1	191.3	23.3	5.9				
		1840	3229	394.8	46.2	396.6	6.1	19.65	0.20	7939.8	203.3	31.4	5.0				
		3500	4454	179.1	21.4	179.2	3.4	20.40	0.20	3666.0	108.9	26.2	5.5				
		4550	5004	115.3	14.3	114.7	2.5	20.59	0.21	2588.1	93.3	18.1	2.7				
		5200	5579	47.8	6.3	47.7	1.3	20.78	0.21	1487.3	70.0	23.4	5.6				
3	Natural Gas	475	1139	264.8	31.9	266.1	4.1	20.12	0.20	5213.2	138.8	20.7	8.7				
		1300	2224	407.7	49.5	397.6	6.1	19.61	0.20	8068.6	258.7	14.9	5.0				
		2555	3449	179.4	22.0	178.1	2.9	20.39	0.20	3692.6	94.7	20.9	5.1				
		3645	4019	115.9	14.3	113.1	2.0	20.59	0.21	2576.8	110.3	9.2	2.9				
		4390	5249	73.8	9.1	74.4	1.8	20.72	0.21	1846.6	67.3	4.0	3.2				
4	Heptane (pool)	1375	2149	153.3	19.0			20.49	0.21	3761.3	208.9	67.2	21.1			43.6	21.6
		2850	3334	268.7	33.0	273.9	5.3	20.14	0.21	6112.6	227.9	132.2	28.6			291.9	87.3
		4090	5489	374.9	45.3	425.1	8.8	19.83	0.22	7456.2	480.7	230.3	77.6			864.6	216.0
5	Heptane (pool)	1245	1799	140.5	17.3			20.55	0.21	3245.6	169.5	40.2	16.4			71.0	24.7
		2340	2969	221.2	27.2			20.33	0.21	4304.1	168.0	152.0	43.1			409.8	105.6
6	Natural Gas	660	1539	73.6	9.2	75.2	1.5	20.77	0.21	1756.1	75.2	2.1	2.4				
		2515	4179	173.7	20.2	179.3	5.1	20.48	0.21	3295.0	123.5	50.0	15.4				
		4425	4944	272.1	32.5	268.6	4.2	20.19	0.21	4922.9	164.7	21.9	3.4				
		5090	5724	417.5	49.5	399.3	9.0	19.72	0.20	7550.3	228.9	18.8	2.7				
		6090	6549	80.5	10.1	75.0	1.6	20.76	0.21	1708.8	59.4	1.7	2.2				
6 5	Natural Gas	285	679	96.5	11.8	76.1	8.8	20.71	0.21	2178.2	76.6	176.6	26.9	7.2	5.1		
		920	1204	423.9	51.9	399.4	6.1	19.64	0.20	8246.5	384.4	111.4	56.9	58.2	12.4		
		1600	2329	272.8	32.2	269.2	12.8	20.15	0.20	5170.4	118.6	45.2	13.8	20.6	9.2		
		2540	2804	181.3	22.3	178.7	10.8	20.45	0.21	3510.4	88.9	31.5	8.1	8.5	5.6		
		2980	3259	85.5	10.7	74.1	10.6	20.75	0.21	1842.6	107.4	55.8	19.3	4.2	5.1		
7	Heptane (pool)	1200	1669	147.7	18.6			20.50	0.21	3714.7	182.5	12.4	2.9	2.2	5.1	52.0	21.9
		2105	2664	246.2	32.2			20.15	0.22	5888.7	421.3	71.5	23.4	11.8	7.2	274.0	90.6
		3040	3709	341.4	41.1	404.5	7.4	19.83	0.21	7465.3	292.3	247.6	47.6	24.2	10.4	820.8	167.3
8	Methanol (pool)	1439	2009	17.2	2.2	22.5	71.4	20.94	0.21	825.9	58.8	0.0	2.0	1.0	5.0		
9	Ethanol (pool)	1300	2019	19.3	2.4	5.8	19.6	20.94	0.21	881.6	48.3	0.0	2.0	1.0	5.0		
10	Toluene (pool)	1400	1884	48.9	6.8	48.8	2.3	20.84	0.21	1623.4	156.2	70.8	37.8	42.2	6.3	115.7	32.5
		2805	3154	137.6	18.1	179.3	3.3	20.51	0.21	4151.4	255.2	95.2	10.1	9.6	5.7	306.5	56.1
		3600	4224	202.2	25.0	270.4	4.6	20.26	0.21	6059.0	352.3	41.4	9.7	5.4	5.4	458.4	105.1
		4435	5044	295.4	37.6	399.9	6.3	19.93	0.21	8512.8	533.6	47.1	22.9	2.0	5.0	486.2	180.3
		5120	5394	338.8	43.5	498.7	7.8	19.93	0.29	8267.7	1495.5	107.2	38.8	1.3	5.1	682.8	217.2
11	Ethanol (spray)	550	1039	82.5	11.4	74.0	2.0	20.72	0.21	2412.6	302.7	117.4	99.9	3.8	5.3		
		1400	1714	143.7	18.1	181.0	3.1	20.53	0.21	3805.4	141.8	12.3	11.6	2.0	5.0		
		2175	2849	263.1	32.1	268.8	4.5	20.11	0.21	6721.1	312.7	19.6	5.2	4.5	5.2		
		2940	4200	335.3	41.2	398.5	6.6	19.84	0.21	8531.0	728.5	54.1	7.9	19.1	6.8		
12	Methanol (spray)	300	724	71.8	9.3	85.0	2.1	20.77	0.21	1977.7	106.0	0.7	2.2	1.0	5.0		
		1145	1609	142.6	17.9	181.0	3.1	20.54	0.21	3585.3	209.5	1.5	2.3	1.0	5.0		
		1949	2669	239.8	28.8	270.5	4.3	20.21	0.21	5890.9	376.1	35.7	7.4	1.0	5.0		
		2760	3299	306.5	37.1	400.6	6.6	19.96	0.21	7612.1	470.1	87.4	15.8	2.0	5.0		
13	Polystyrene	710	1344	14.9	2.0	0.0	1.0	20.93	0.21	845.0	57.5	17.5	2.9	9.5	5.1	47.5	23.0
14	Polystyrene	870	1724	67.3	8.8	0.0	1.0	20.76	0.21	2119.2	180.2	80.3	10.6	36.6	7.2	304.3	173.6
15	Heptane (spray)	280	759	87.7	12.2	82.0	22.8	20.69	0.21	2406.0	198.3	6.4	2.4	2.0	5.0	31.4	21.8
		950	1259	160.2	19.7	179.8	5.9	20.43	0.21	4149.7	124.1	20.0	4.0	2.6	5.1	111.9	25.9
		1475	1999	227.3	28.1	271.8	4.4	20.20	0.20	5485.7	212.1	63.0	17.2	9.1	6.3	328.0	96.0
		2200	2764	300.8	39.1	401.2	6.2	19.92	0.21	6789.6	256.9	243.8	31.6	41.1	15.8	912.8	166.1
		2790	3169	377.3	50.6	501.5	7.7	19.63	0.20	8412.1	317.1	384.1	35.9	89.4	22.5	1155.3	162.4
		3390	3734	83.3	11.9	84.7	5.1	20.69	0.21	2379.4	159.0	9.7	3.3	2.0	5.0	45.1	24.9
16	Polystyrene	545	649	358.1	45.7	0.0	1.0	19.73	0.20	9108.1	214.6	251.1	30.3	3.0	5.0	1258.4	228.3
		715	769	308.7	40.0	0.0	1.0	19.86	0.20	8814.7	213.8	78.2	38.4	2.0	5.0	537.1	183.6

3.2 Temperatures

The time history of the interior gas temperature is shown at 4 locations during the natural gas test (#3) in Fig. 30. This figure shows the difference in temperature between the front and rear gas sample locations and the temporal variation of temperature in the compartment. Refer to Table 2 and Fig. 4 for exact locations of the temperature probes. The measurement labels for the figures and tables in this section are described in Table 11. The aspirated thermocouple pumps remained on for the duration of this test. The general trend for these tests was higher temperatures in the upper layer at the front sample location than at the rear, however the magnitude of this difference was a function of fuel type and fire size. For all of the natural gas fires, the front and rear gas temperatures in the lower layer of the enclosure (24 cm from the floor) were not significantly different. The front to rear variation in lower layer temperature was more pronounced for the other fuels.

Table 11. Description of interior gas temperature measurement labels.

Measurement Label	Description
15 TRSampA (C)	Aspirated thermocouple at rear sample location (88 cm above floor)
16 TFSampA (C)	Aspirated thermocouple at front sample location (88 cm above floor)
17 TR24A (C)	Aspirated thermocouple at lower rear location (24 cm above floor)
18 TR80A (C)	Aspirated thermocouple at upper rear location (80 cm above floor)
19 TF24A (C)	Aspirated thermocouple at lower front location (24 cm above floor)
20 TF80A (C)	Aspirated thermocouple at upper front location (80 cm above floor)

For a number of the tests, the aspiration flow pumps were run intermittently. This was done to observe the effect of the aspiration flow on the temperature and other measurements at nearby locations, as well as to conserve the water traps and filters. It was determined that turning on or off the suction flow to the aspirated thermocouples had no measurable effect on the observed gas species volume fractions. This is an important result since the separation distance between the probes was less than 3 cm. Further analysis of probe interactions can be found in Appendix B. The thermocouple response to cycling the aspiration flow is show in Fig. 31. This figure shows the front aspirated thermocouple measurements at two different heights in the enclosure for test #15 using heptane and the spray burner. The HRR was approximately 220 kW during this time window. The upper series in Fig. 31 is the aspirated thermocouple at the front sample location (TFSampA in Table 2), 10 cm below the ceiling. Because this probe was in a region with high soot mass fraction, the aspiration had little effect on the average temperature results. The faster time response (due to high convective heat transfer while aspirating) increases the measurement variation. The high soot mass fraction caused little difference to the probe's average temperature with and without aspiration because the probe's optical view of any cooler temperature radiative heat sinks was minimal due to soot blockage.

The aspirated temperature probe located inside the front of the enclosure at a height of 24 cm above the floor (TF24A) is also shown. The measured temperature at this location decreased by more than 500 °C when the aspiration was applied. This result shows the important application of aspirated thermocouples to the thermal characterization of flashed-over compartment fires. Although differences between bare-beads and aspirated thermocouples are usually much greater in the lower layer, the upper layer differences may be as much as 100 °C to 200 °C depending, primarily, on a fuel's sooting characteristics.

A detailed analysis of the aspirated thermocouple time response and uncertainty can be found in Appendix A of this report. All of the temperature results reported here represent the temperature of the thermocouple bead, not the true temperature of the gas.

Average temperatures were calculated over pseudo-steady periods for all of the tests. Temperature measurements collected when the aspiration pumps were off or during a transient period were not included in the averages. A summary of the rear gas temperature measurements with combined expanded uncertainty (U) are listed in Table 12. The front gas temperatures are given in Table 13. The average temperatures are plotted as a function of HRR for the natural gas full-door tests in Fig. 32. These temperature measurements demonstrate the reproducibility of the measurements over a number of different days. Figure 33 and Fig. 34 show the steady temperatures at the front and rear gas sample location for all of the fuels included in this study. In general, the soot producing fires (heptane, toluene, polystyrene) produced hotter gas temperatures inside the enclosure than the cleaner fires (natural gas, alcohols) at the same measured HRR.

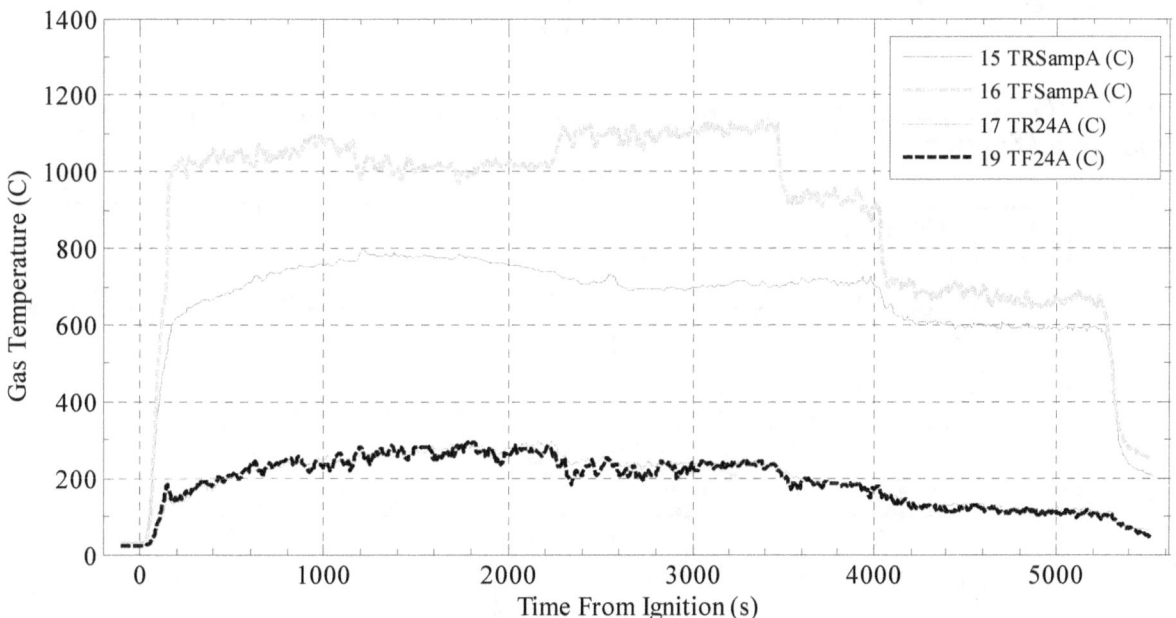

Figure 30. Gas temperature measurement results for 4 positions inside RSE. Natural gas test #3.

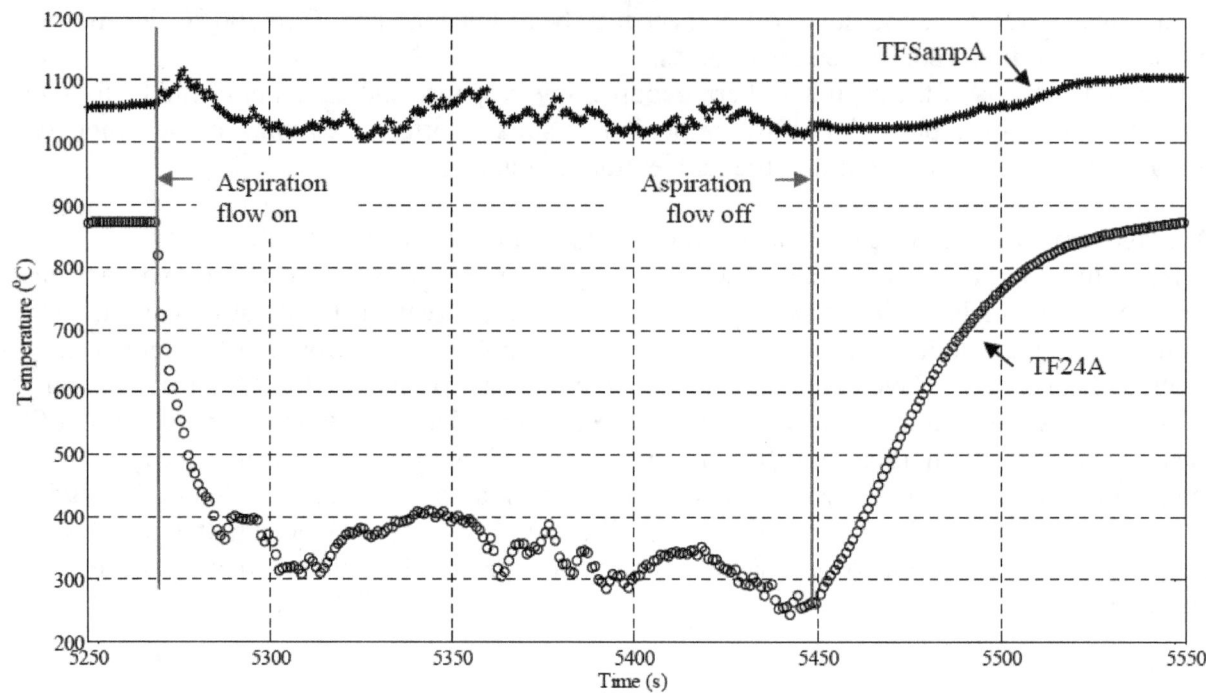

Figure 31. Gas temperature measurements at two front interior locations inside RSE during the heptane spray fire (test # 15).

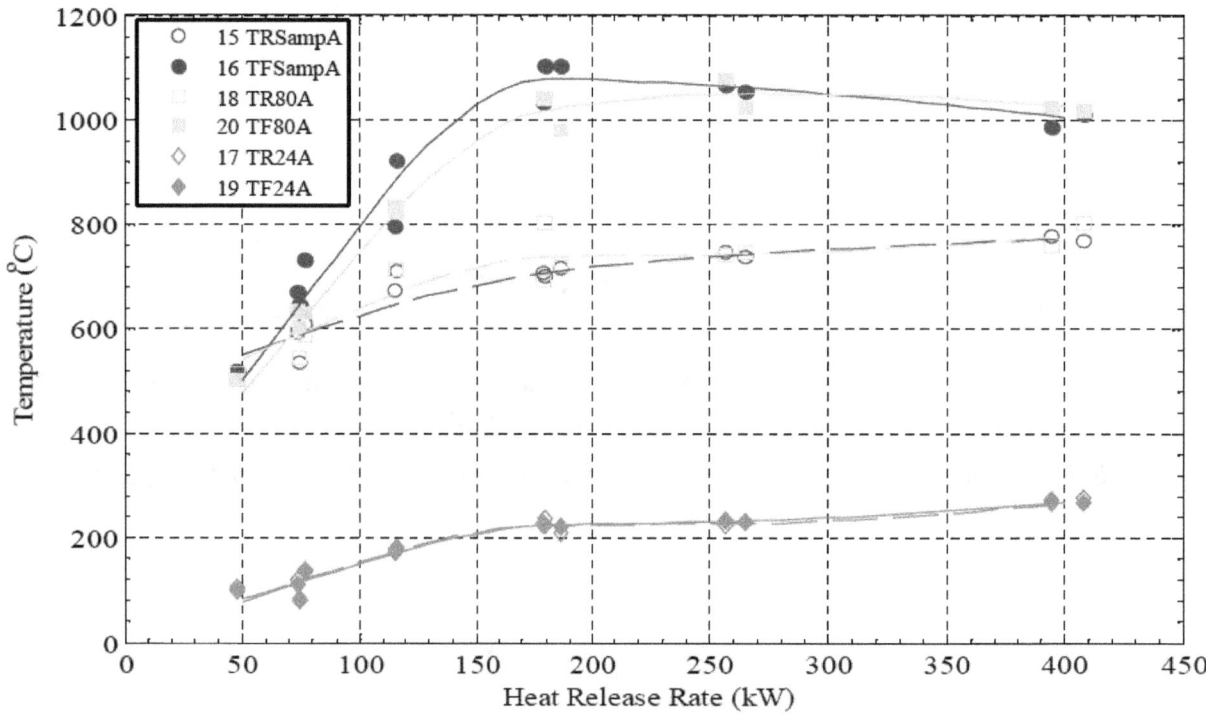

Figure 32. Steady state average temperature measurements at interior locations for repeated natural gas fires (tests #1, #2, #3). The lines in this figure are piecewise cubic polynomial fits to the data.

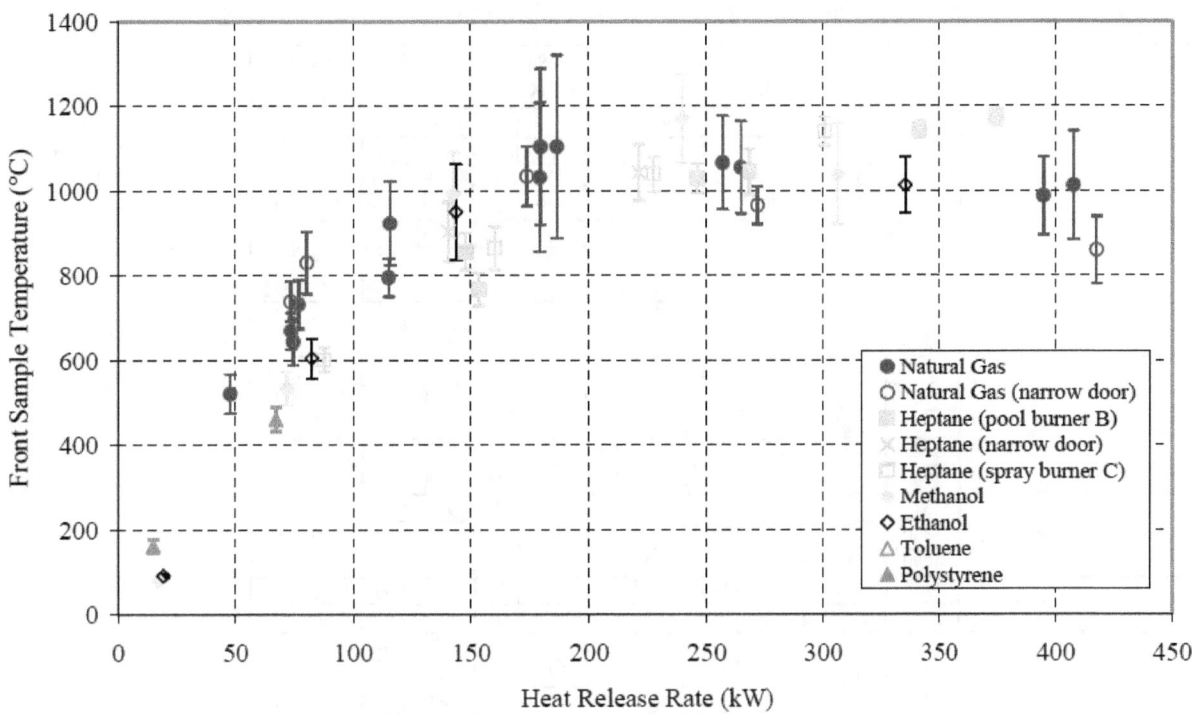

Figure 33. Steady state average temperature results from aspirated thermocouple measurement at front gas sampling location. **note:** Front sample thermocouple failed during toluene fire (test #10) and polystyrene fire (test #16)

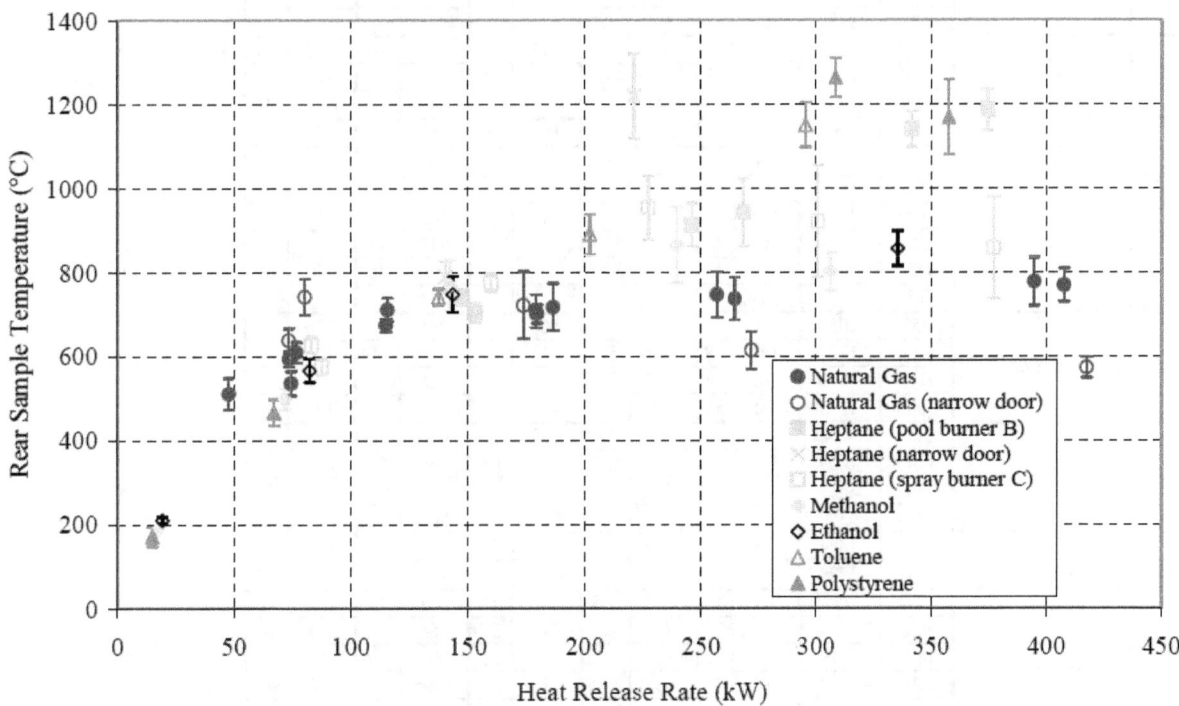

Figure 34 Steady state average temperature results from aspirated thermocouple measurement at rear gas sampling location.

Table 12. Steady state rear gas temperatures and total heat flux to the floor inside the RSE.

Test #	Fuel	SS Window		HRR (kW)		HFR (kW/m^2)		TRSampA (C)		TR80A (C)		TR24A (C)	
		start (s)	stop (s)	Mean	U	Mean	U	Mean	U	Mean	U	Mean	U
1	Natural Gas	500	1924	75	11	17.1	4.1	536	27	546	32	80	28
		2220	3384	186	23	57.9	10.8	717	53	719	46	211	25
		3600	3944	77	11	30.2	3.6	609	20	589	29	140	16
2	Natural Gas	400	1594	257	32	57.6	18.0	747	58	740	69	225	44
		1840	3229	395	49	75.6	6.3	777	56	759	37	268	11
		3500	4454	179	23	62.2	4.6	708	46	691	40	228	11
		4550	5004	115	15	43.3	4.1	674	11	715	25	178	12
		5200	5579	48	7	18.1	2.9	511	37	513	19	105	22
3	Natural Gas	475	1139	265	33	37.8	9.0	737	53	747	50	232	30
		1300	2224	408	55	55.3	7.8	769	34	803	33	277	17
		2555	3449	179	25	50.0	3.4	701	19	803	48	237	10
		3645	4019	116	14	35.3	2.8	712	23	715	26	182	14
		4390	5249	74	10	20.2	2.1	594	16	598	18	121	15
4	Heptane (pool)	1375	2149	153	21	52.9	7.0	703	22	707	26	234	22
		2850	3334	269	34	96.9	44.1	942	82	939	61	430	66
		4090	5489	375	54	235.0	35.4	1186	50	1171	43	845	274
5	Heptane (pool)	1245	1799	140	18	67.6	9.9	789	36	783	33	441	64
		2340	2969	221	29	220.7	34.2	1219	101	1248	90	1254	170
6	Natural Gas	660	1539	74	11	24.3	4.3	638	28	644	30	186	28
		2515	4179	174	22	70.6	5.9	722	80	781	44	390	23
		4425	4944	272	33	76.5	5.2	614	44	698	36	418	15
		5090	5724	417	50	73.9	4.8	572	25	654	28	420	14
		6090	6549	80	11	42.8	5.0	742	42	738	35	285	22
6.5	Natural Gas	285	679	97	12	23.8	3.0						
		920	1204	424	51	83.1	12.0						
		1600	2329	273	33	87.9	7.4						
		2540	2804	181	22	77.4	6.9						
		2980	3259	85	11	29.5	4.3						
7	Heptane (pool)	1200	1669	148	20	54.1	6.5	741	21	718	16	210	14
		2105	2664	246	38	133.0	41.5	913	59	921	54	379	34
		3040	3709	341	43	204.8	37.0	1139	45	1125	32	642	94
8	Methanol (pool)	1439	2009	17	3	2.4	0.5	188	6	185	8	67	5
9	Ethanol (pool)	1300	2019	19	3	2.7	0.4	210	9	203	9	75	5
10	Toluene (pool)	1400	1884	49	9	14.6	2.4						
		2805	3154	138	20	65.7	7.0	741	21	763	21	240	20
		3600	4224	202	27	116.1	15.7	890	53	919	45	354	36
		4435	5044	295	43	257.4	37.3	1150	58	1254	56	970	89
		5120	5394	339	44	336.2	42.9						
11	Ethanol (spray)	550	1039	83	15	19.9	2.5	566	26	546	20	126	6
		1400	1714	144	18	44.6	5.0	748	44	729	36	221	15
		2175	2849	263	34	80.5	8.1			848	44	307	13
		2940	4200	335	51	80.9	5.9	856	45			305	10
12	Methanol (spray)	300	724	72	10	13.4	2.4	499	26	490	27	94	11
		1145	1609	143	19	39.5	5.7	769	64	736	51	192	20
		1949	2669	240	30	62.4	4.3	864	95	842	105	259	13
		2760	3299	306	38	61.1	4.2	801	50	803	45	260	13
13	Polystyrene	710	1344	15	3	3.5	0.9	171	23	183	26	112	19
14	Polystyrene	870	1724	67	12	23.6	5.1	467	30	467	26	130	13
15	Heptane (spray)	280	759	88	16	30.8	6.5	575	18	571	15	155	7
		950	1259	160	19	76.9	8.0	775	27	784	21	325	14
		1475	1999	227	30	134.3	19.2	954	83	999	106	497	85
		2200	2764	301	46	143.0	20.5	921	140	1005	137	468	42
		2790	3169	377	58	135.7	11.8	858	129	901	127	452	27
		3390	3734	83	15	46.7	11.0	627	22	628	17	219	22
16	Polystyrene	545	649	358	42	104.2	20.5	1169	84	1240	130	1256	137
		715	769	309	35	169.9	17.5	1263	36	1160	45		

46

Table 13. Steady state front gas temperatures and total heat flux to the floor inside the RSE.

Test #	Fuel	SS Window		HRR (kW)		HFF (kW/m2)		TFSampA (C)		TF80A (C)		TF24A (C)	
		start (s)	stop (s)	Mean	U	Mean	U	Mean	U	Mean	U	Mean	U
1	Natural Gas	500	1924	75	11	19.7	4.6	645	53	604	48	85	35
		2220	3384	186	23	69.4	10.1	1104	206	979	136	223	21
		3600	3944	77	11	28.0	6.2	733	53	631	34	137	20
2	Natural Gas	400	1594	257	32	80.6	16.2	1066	101	1075	108	235	38
		1840	3229	395	49	94.8	7.6	988	84	1025	110	274	23
		3500	4454	179	23	79.6	5.7	1032	173	1043	85	226	23
		4550	5004	115	15	52.7	5.2	796	39	832	61	173	17
		5200	5579	48	7	20.5	3.2	520	46	503	23	101	21
3	Natural Gas	475	1139	265	33	80.6	12.8	1055	99	1024	80	232	32
		1300	2224	408	55	95.8	7.8	1013	119	1016	140	269	27
		2555	3449	179	25	80.4	5.7	1104	175	1038	142	226	28
		3645	4019	116	14	52.8	4.2	924	91	823	62	179	19
		4390	5249	74	10	28.3	3.0	670	40	633	27	113	16
4	Heptane (pool)	1375	2149	153	21	51.3	5.1	767	39	847	44	284	12
		2850	3334	269	34	98.4	11.8	1044	54	1090	77	468	58
		4090	5489	375	54	168.1	21.9	1173	21	1210	33	677	97
5	Heptane (pool)	1245	1799	140	18	80.8	12.0	903	71	991	70	569	63
		2340	2969	221	29	119.2	16.5	1043	66	1149	50	1037	144
6	Natural Gas	660	1539	74	11	9.8	1.6	739	48	714	44	210	30
		2515	4179	174	22	43.2	7.6	1035	71	1032	84	423	14
		4425	4944	272	33	51.5	3.6	966	45	978	45	456	14
		5090	5724	417	50	50.2	3.3	860	79	829	63	450	10
		6090	6549	80	11	28.9	4.1	830	73	804	43	302	24
6.5	Natural Gas	285	679	97	12	24.7	2.7						
		920	1204	424	51	88.8	8.9						
		1600	2329	273	33	90.9	6.6						
		2540	2804	181	22	77.2	5.5						
		2980	3259	85	11	29.1	3.8						
7	Heptane (pool)	1200	1669	148	20	57.8	6.4	857	39	831	28	276	19
		2105	2664	246	38	118.5	12.2	1029	35	1091	41	423	40
		3040	3709	341	43	166.7	13.4	1145	22	1230	56	594	63
8	Methanol (pool)	1439	2009	17	3	2.2	0.4	80	5	181	8	62	5
9	Ethanol (pool)	1300	2019	19	3	2.6	0.4	91	6	198	8	76	6
10	Toluene (pool)	1400	1884	49	9	13.6	2.4						
		2805	3154	138	20	59.2	6.6			859	27	297	24
		3600	4224	202	27	101.9	10.3			1067	54	421	50
		4435	5044	295	43	174.8	17.1			1241	38	643	75
		5120	5394	339	44	210.0	22.2						
11	Ethanol (spray)	550	1039	83	15	17.6	3.3	605	46	541	27	134	11
		1400	1714	144	18	45.3	5.9	951	108	851	70	224	39
		2175	2849	263	34	86.8	8.9			1085	77	308	23
		2940	4200	335	51	82.3	6.4	1013	59			288	25
12	Methanol (spray)	300	724	72	10	11.7	2.6	533	40	493	41	100	14
		1145	1609	143	19	40.0	5.9	993	95	850	109	200	22
		1949	2669	240	30	64.9	4.9	1171	94	1134	109	252	32
		2760	3299	306	38	61.5	4.1	1040	113	1144	122	244	33
13	Polystyrene	710	1344	15	3	2.3	0.6	161	17	167	21	88	17
14	Polystyrene	870	1724	67	12	17.2	4.5	460	28	454	26	139	19
15	Heptane (spray)	280	759	88	16	26.1	7.1	601	29	569	26	157	11
		950	1259	160	19	74.2	8.9	864	50	870	22	309	21
		1475	1999	227	30	128.5	13.7	1040	47	1121	43	346	71
		2200	2764	301	46	145.0	12.8	1140	38	1197	44	382	95
		2790	3169	377	58	144.0	11.7			1169	36	405	81
		3390	3734	83	15	38.2	8.4			629	21	161	57
16	Polystyrene	545	649	358	42	71.2	14.7			1195	56	968	111
		715	769	309	35	90.7	6.7			1244	29	947	64

3.3 Doorway Velocities and Temperatures

Doorway velocity measurements were performed in order to establish the ventilation conditions of the compartment fires. Velocity probes and thermocouples were placed at five different heights in the doorway. Aspirated thermocouples and bare bead thermocouples were placed half-way between the velocity probes at three different heights to get a first order estimate of the error in temperature with the bare bead thermocouples due to radiation. See Fig. 4 and Table 3 for exact locations of the doorway probes.

Figure 35 compares the bare bead and aspirated thermocouple temperature measurements at three different heights in the doorway for all 17 fire tests. The results show that at heights of 50 cm and 70 cm in the doorway the bare bead temperate measurements agree reasonably well with the aspirated temperature measurements. This was because these points were above the neutral plane where the velocities were relatively high and the large opacity of the hot gases in the upper layer reduced the radiation losses. However, at 30 cm, the temperature probes are below the neutral plane where the velocity is lower and the gas is optically thin cool room air. This thermocouple is exposed to radiation from the hot upper layer resulting in a significant overestimate of the gas temperature. The results of this analysis are that the error in velocity due to using the bare-bead temperature is less than 10 % above the neutral plane and less than 30 % below the neutral plane. Figure 36 show the doorway temperature measurements (not corrected for radiation) for a steady natural gas fire at 265 kW during test #3. The error bars in this figure represent twice the standard deviation of the measurement and does not include uncertainty due to radiation.

The steady state velocity profiles along the vertical centerline of the doorway for test #3 using natural gas are shown in Fig. 37. As expected, the magnitude of the velocity measurement increases and the location of the neutral plane moves downward as the fire size increases. Figure 38 shows the steady doorway velocity profiles for test #15 using the heptane spray burner. The lines shown in both of these figures are drawn to highlight trends in the data and do not represent a physical model.

The mass flow rate of air into the enclosure is often used to define a global equivalence ratio. This value can be found by integrating the product of velocity and density over the area below the neutral plane. Although this calculation was attempted, the uncertainty in velocity, temperature, and location of neutral plane (due to sparse data) prevented a meaningful determination of mass flow rate. Future work is planned to better quantify the doorway mass flow.

48

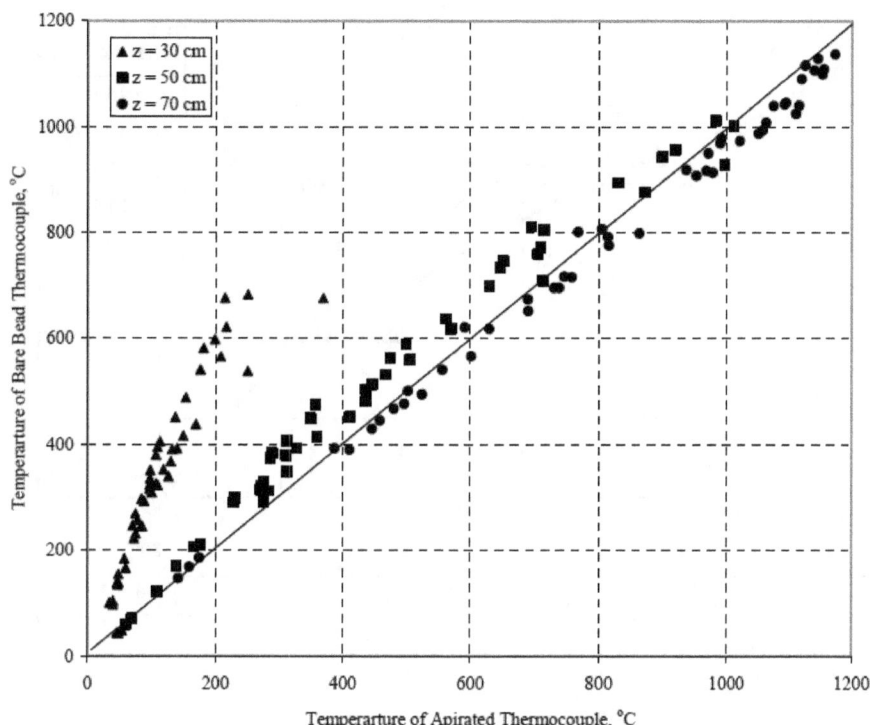

Figure 35. Comparison of doorway centerline temperatures measured using aspirated and bare bead thermocouple at the same position. Steady state average values for all RSE tests.

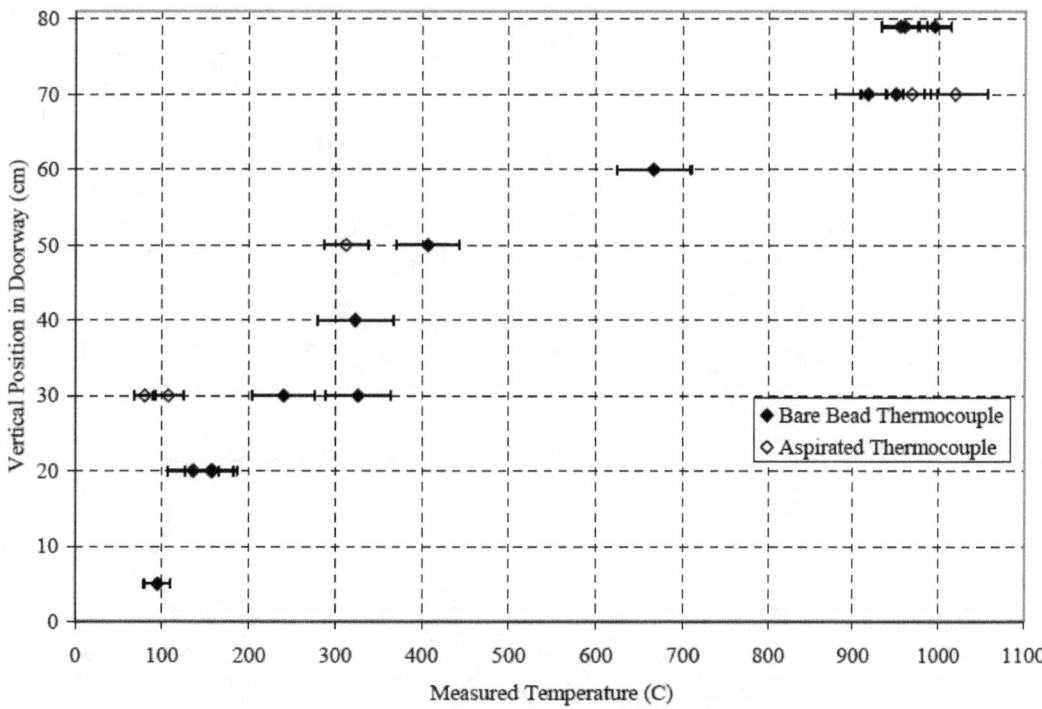

Figure 36. Steady state doorway temperature measurements for natural gas test #3, HRR=265 kW. Error bars in this figure are 2 standard deviations.

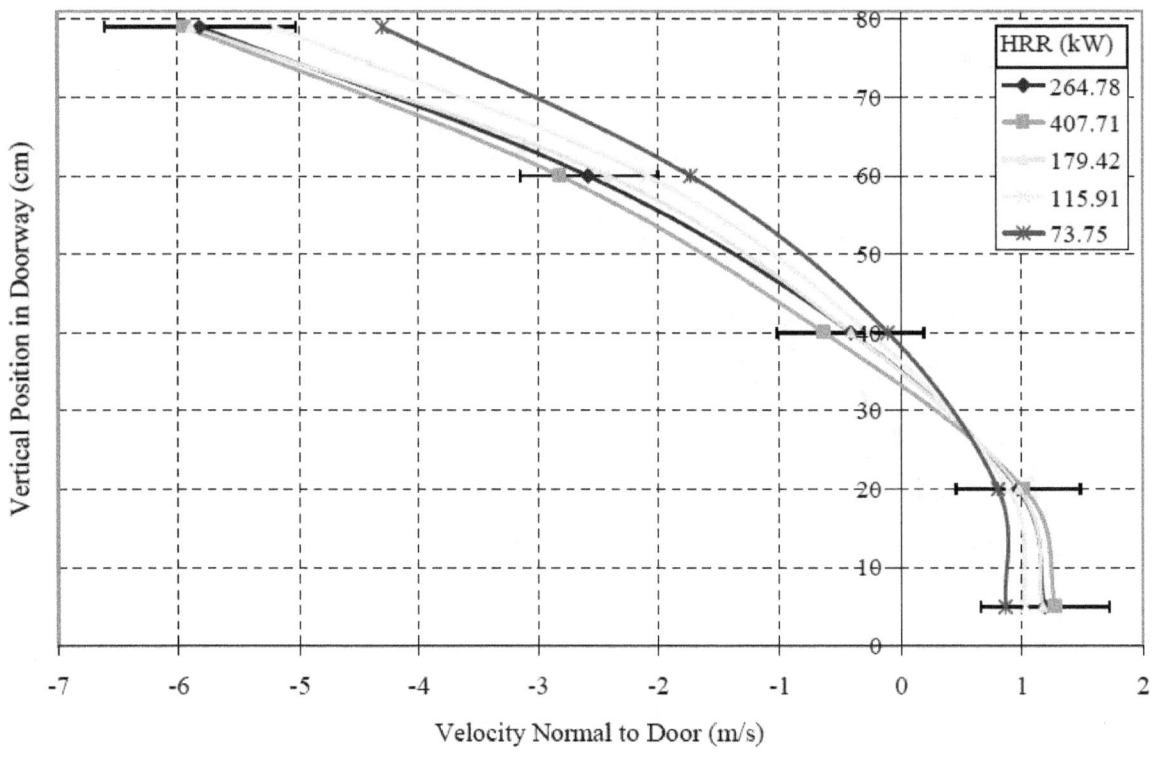

Figure 37. Doorway centerline velocity profiles for natural gas fire (test #3).

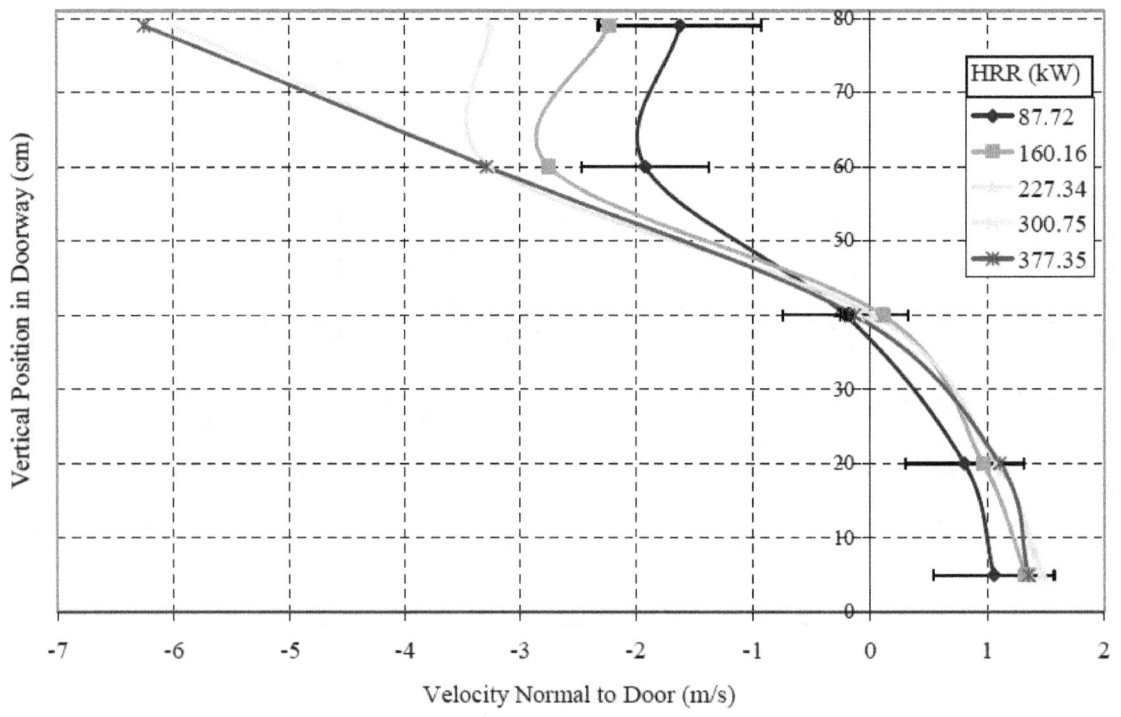

Figure 38. Doorway centerline velocity profiles for heptane spray fire (test #15).

3.4 Interior Gas Species and Soot

The measurement labels for the figures and tables in this section are described in Table 14. The time history of the gas and soot species measurements at the front and rear probe locations for the heptane spray fire (test #15) is shown in Fig. 39. The mean measured HRR value over a given steady time window is annotated on this figure. Several observations of this plot are noteworthy. When the fire was underventilated (between 2000 s and 3000 s) there was a mirroring of the random temporal variations in the CO and CO_2 volume fractions. Also evident in Fig. 39 was a strong positive correlation between CO and total hydrocarbons. The measured gas species volume fractions became less uniform (front to back) as the fire size increased. The uniformity of the upper gas layer was dependant on fire size and fuel type. The transient gas volume fractions and soot mass fractions for polystyrene (test #16) are shown in Fig. 40. Very high soot mass fractions (nearly 11 % at the front sample location) were observed during this test. The total hydrocarbons measured less than 0.1 % for the entire test. This was a surprising result since the oxygen in the upper layer was completely depleted and flames were observed exiting the doorway (see photograph in Fig. 22). More work is needed to understand this result.

Table 14. Description of interior gas species and soot measurement labels.

Measurement Label	Description
2 O2Rear	Rear O_2 volume fraction corrected for water (wet)
3 CO2Rear	Rear CO_2 volume fraction corrected for water (wet)
4 CORear	Rear CO volume fraction corrected for water (wet)
5 THCRear	Rear Total Hydrocarbons volume fraction corrected for water (wet)
6 SootRear	Rear Soot Mass fraction corrected for water (wet)
7 O2Front	Front O_2 volume fraction corrected for water (wet)
8 CO2Front	Front CO_2 volume fraction corrected for water (wet)
9 COFront	Front CO volume fraction corrected for water (wet)
10 THCFront	Front Total Hydrocarbons volume fraction corrected for water (wet)
11 SootFront	Front Soot Mass fraction (g/g), corrected for water (wet)

Figure 41 shows the time averaged species volume fractions as a function of heat release rate for all of the natural gas tests with the full-door configuration (tests #1, #2, #3 and #6.5). The trend lines are included in this figure to help visualize general trends in the data, but do not have a theoretical basis. The figure demonstrates the excellent reproducibility of the gas species measurements and lack of sensitivity of the results to the two different wall lining materials and burners (see Table 1) used in this study.

Figure 42 shows the oxygen volume fraction at the front sample location as a function of heat release rate for the six different fuel types included in this study. Oxygen was depleted for fires larger than about 280 kW. There was some small difference for the various fuel types, with the natural gas fires exhibiting oxygen depletion for slightly smaller values (about 260 kW). The rear sample oxygen measurements are summarized in Fig. 43. For the natural gas fires, the oxygen was depleted in the rear of the enclosure at a lower HRR (≈ 180 kW) than in the front (≈ 250 kW), however this result could not be generalized for all fuels. For example, the toluene and heptane fires displayed significant amounts oxygen at the rear sample location at HRR's where oxygen was completely depleted at the front sample location. This was an unexpected result since flames were observed exiting the doorway (see right side of Fig. 28); however it was not inconsistent with the other species measurements that showed locally lean conditions at the upper rear sample location. In addition, preliminary FDS modeling results showed a similar structure in the compartment. The condition at which oxygen is depleted inside of the compartment is a critical point in the characterization of compartment fire chemistry, indicating when the fire becomes underventilated. In terms of the global equivalence ratio (GER) concept (see Section 1.1 of this report), it is the point at which the GER value is equal to 1.0. Based on the results shown in Figure 42, oxygen depletion could be expected in an ISO 9705 enclosure at about 1800 kW, based on the ventilation scaling relation given in Table 25.

Figure 44 through Fig. 51 shows the steady gas and soot sample results for all of the different fuels at the front and rear locations. The steady state carbon monoxide results are summarized in Fig. 46 (front) and Fig. 47 (rear). As expected, the measured CO values were significantly increased after the fire reached a ventilation limited regime (as indicted by depleted oxygen at the sample locations). The results of the total hydrocarbon volume fraction measurements are shown in Fig. 48 and Fig. 49. Figure 50 and Fig. 51 show the results of the gravimetric soot mass fraction measurements. The species volume fraction results are examined further in Sec. 5 of this report.

The time averaged values for all of the gas and soot species volume fractions are listed in Table 15 and Table 16 for the front and rear sample locations, respectively. The values listed in these tables were used to generate the plots shown in Fig. 41 through Fig. 51. The combined expanded uncertainties listed in these tables represent the absolute percentage (not relative to the mean fractional value).

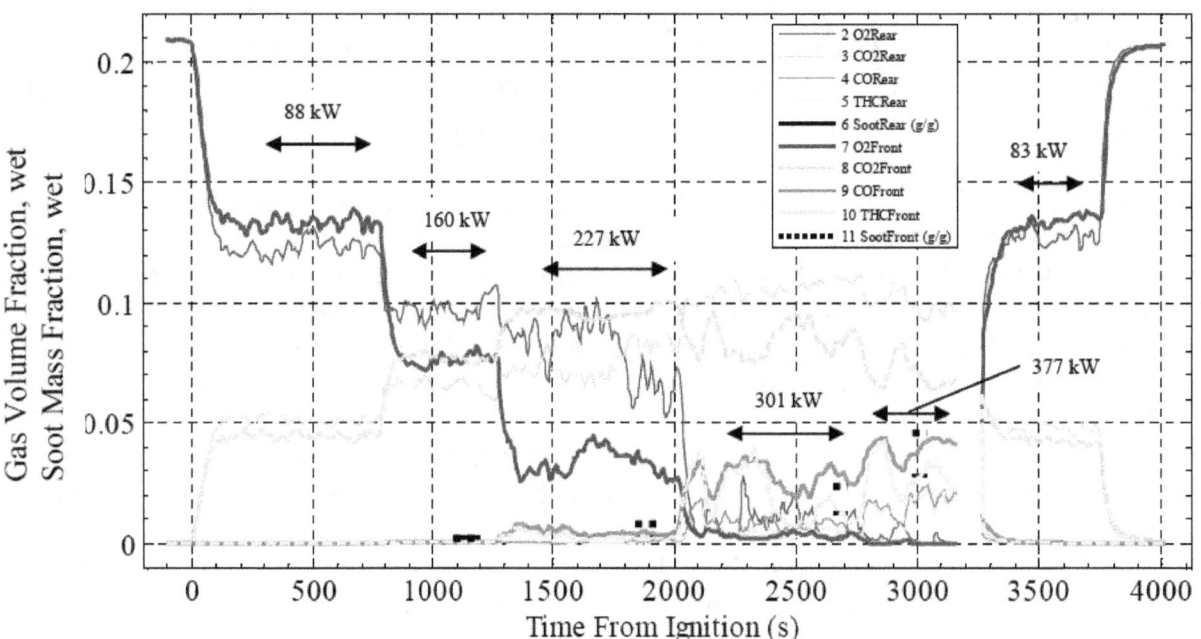

Figure 39. Transient gas volume fractions and soot mass fractions for heptane test #15.

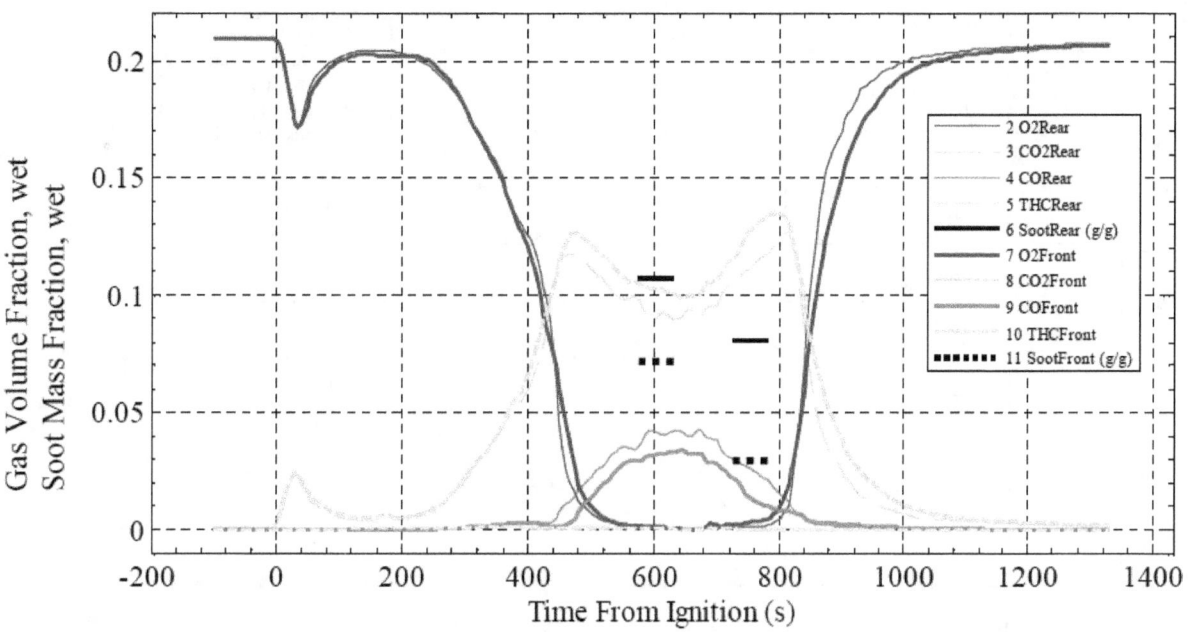

Figure 40. Transient gas volume fractions and soot mass fractions for polystyrene test #16.

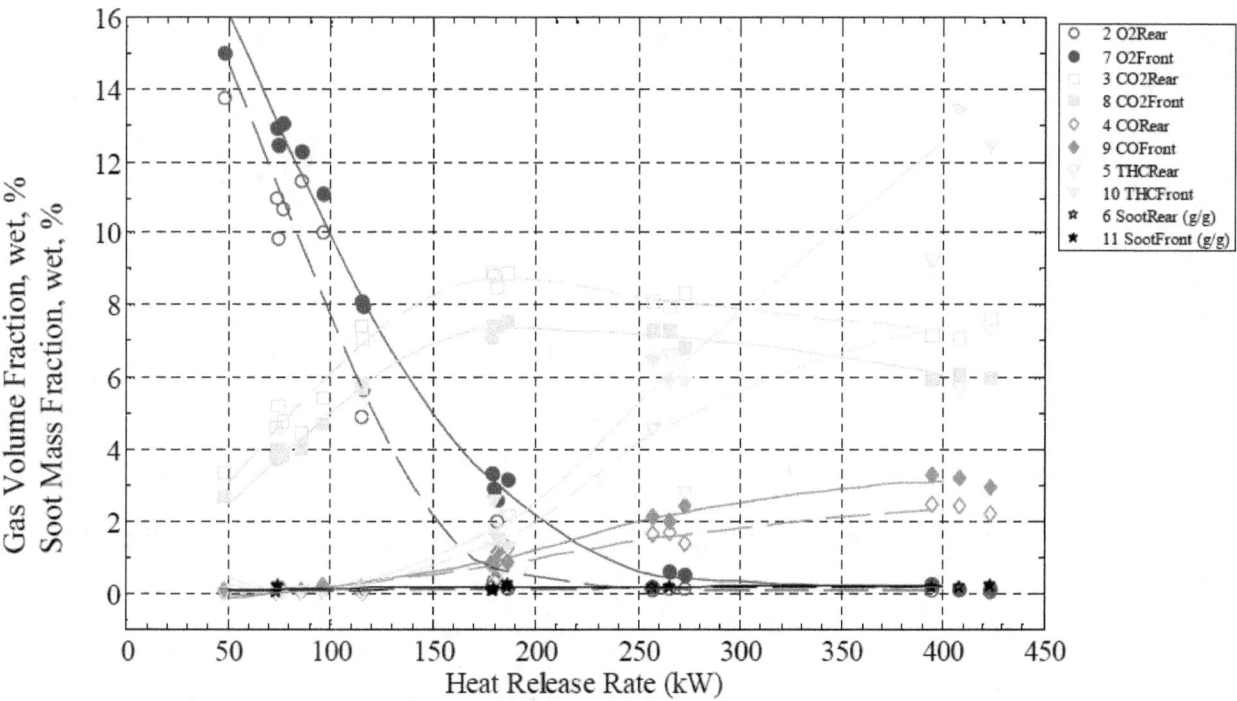

Figure 41. Steady state gas and soot species results for natural gas full door tests #1, #2, #3, and #65. The lines in this figure are piecewise cubic polynomial fits to the data.

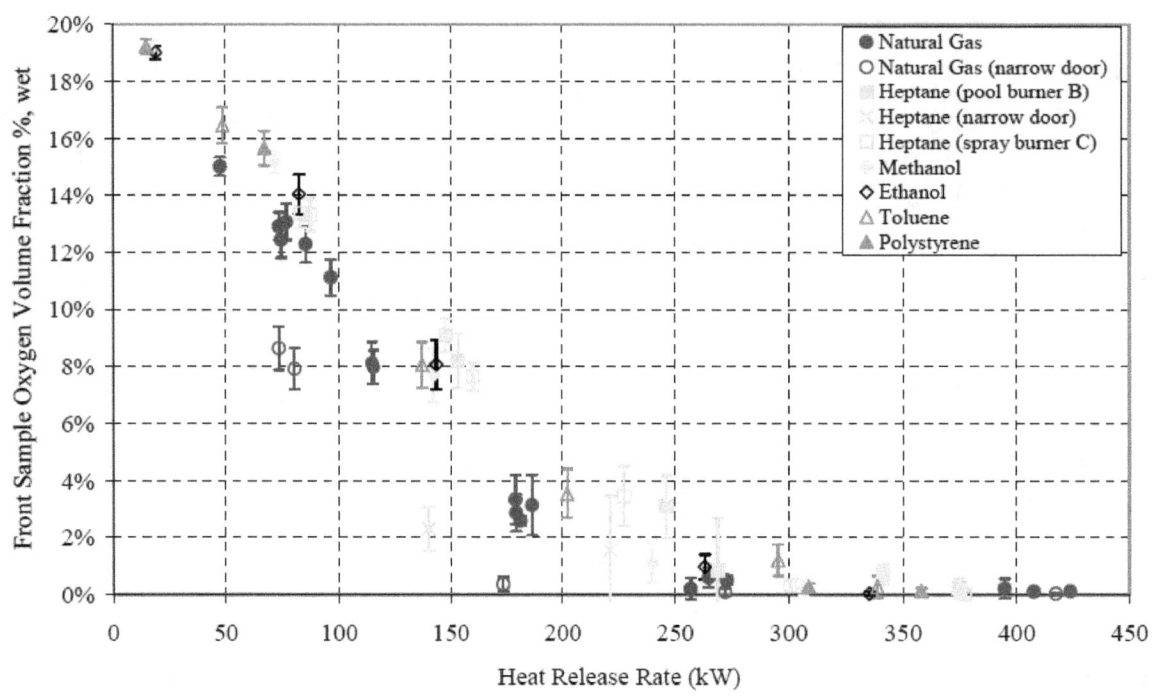

Figure 42. Steady state average oxygen volume fraction measurements at front sample probe location.

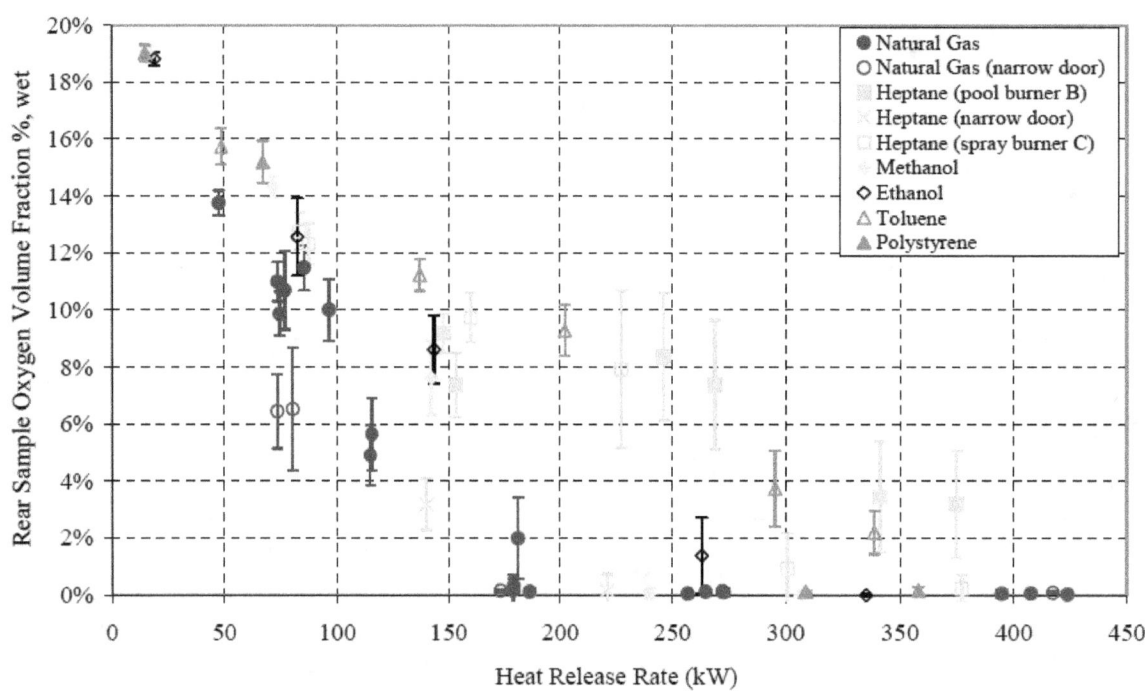

Figure 43. Steady state average oxygen volume fraction measurements at rear sample probe location.

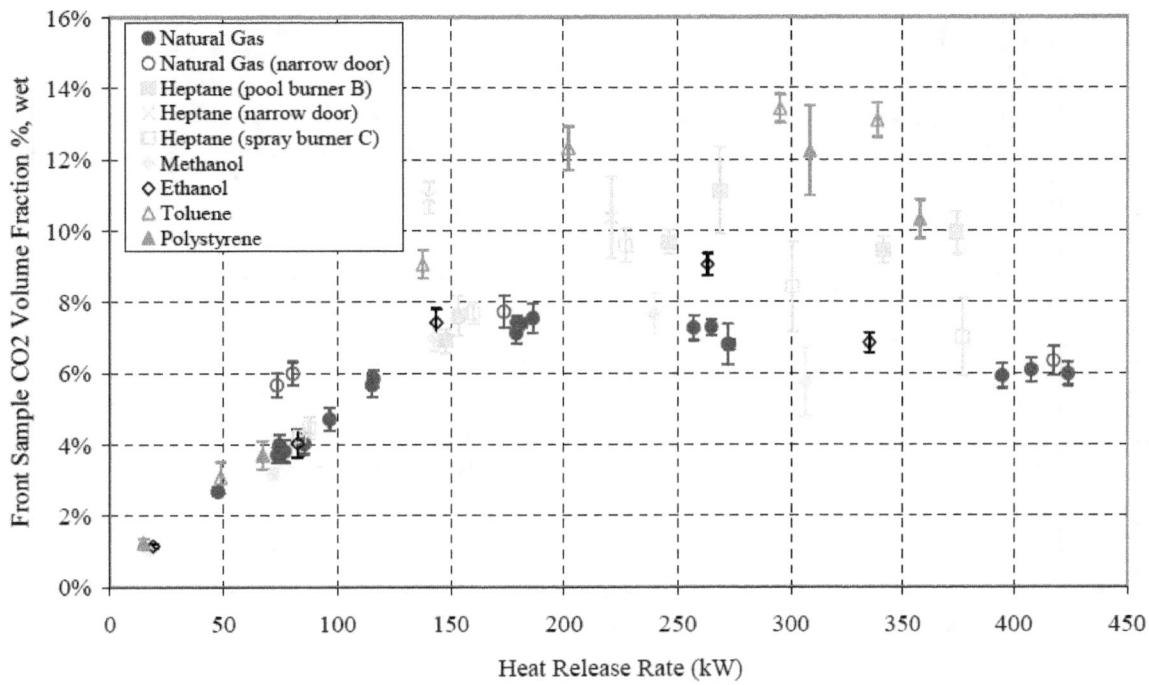

Figure 44. Steady state average carbon dioxide volume fraction measurements at front sample probe location.

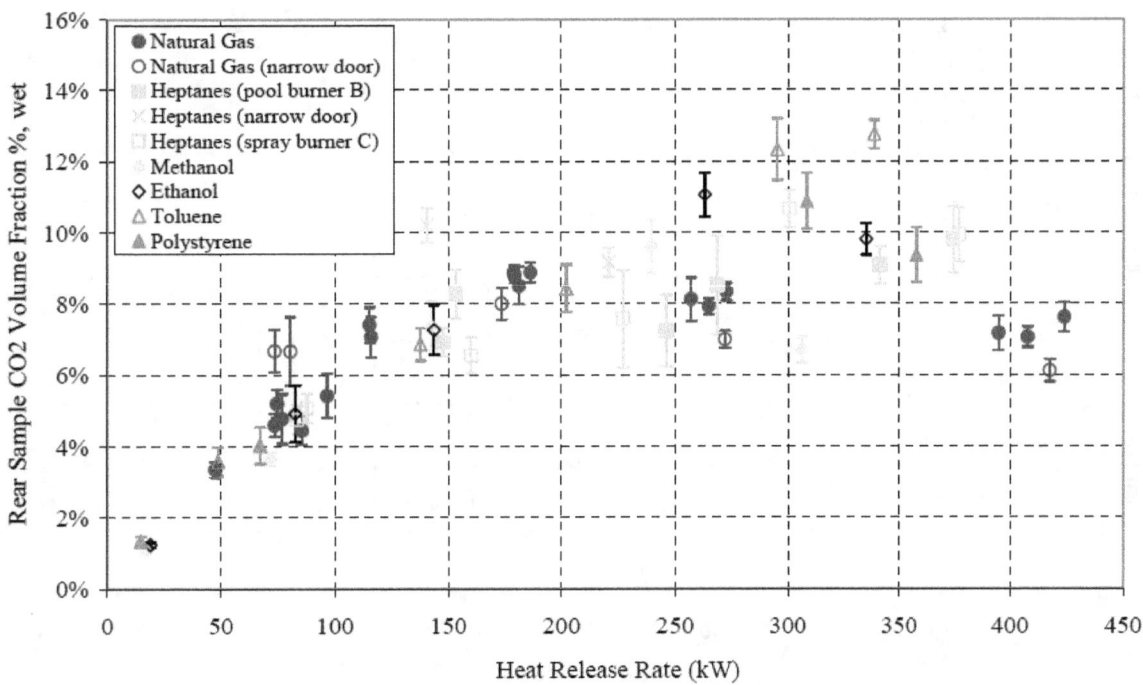

Figure 45. Steady state average carbon dioxide volume fraction measurements at rear sample probe location.

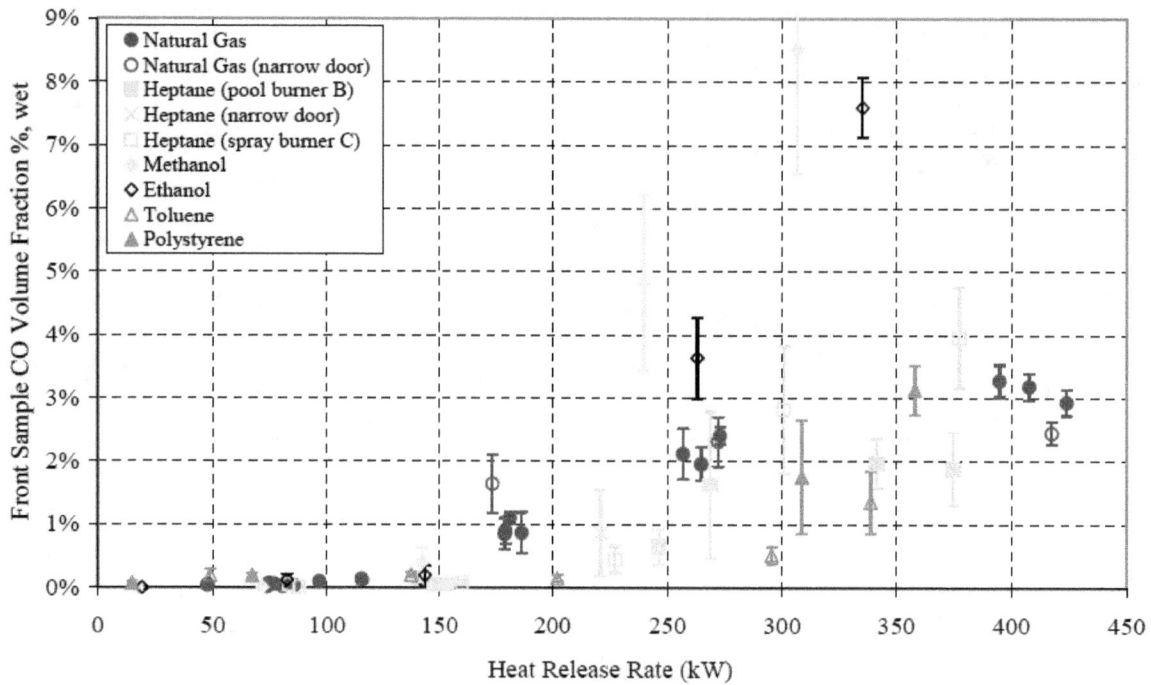

Figure 46. Steady state average carbon monoxide volume fraction measurements at front sample probe location.

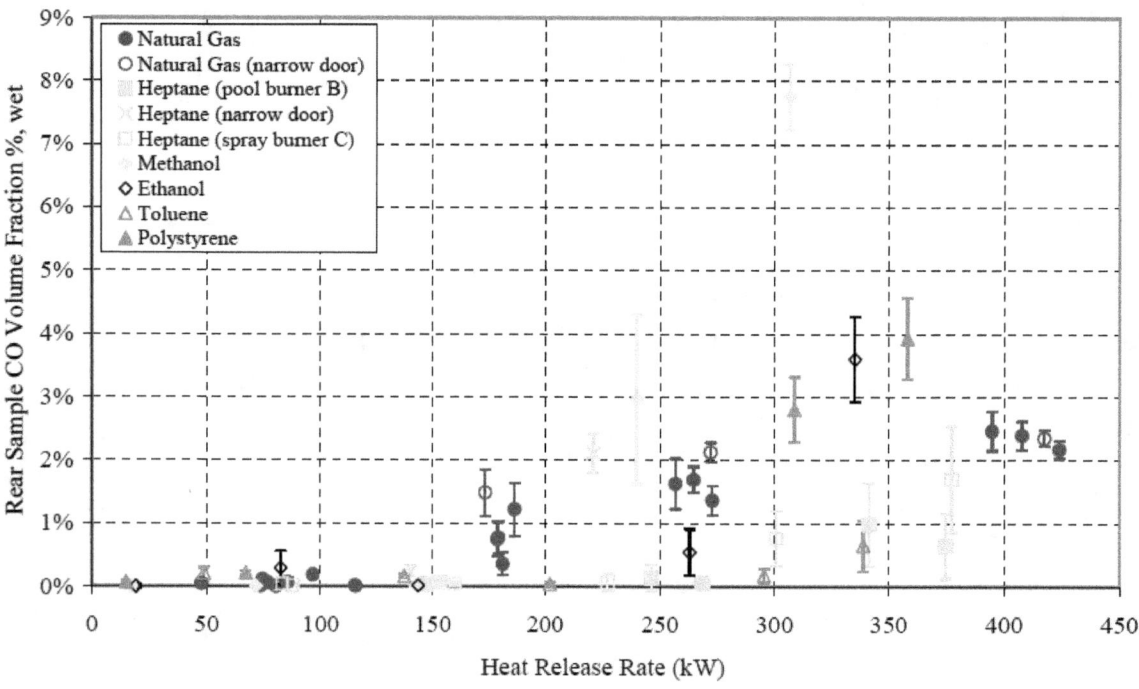

Figure 47. Steady state average carbon monoxide volume fraction measurements at rear sample probe location.

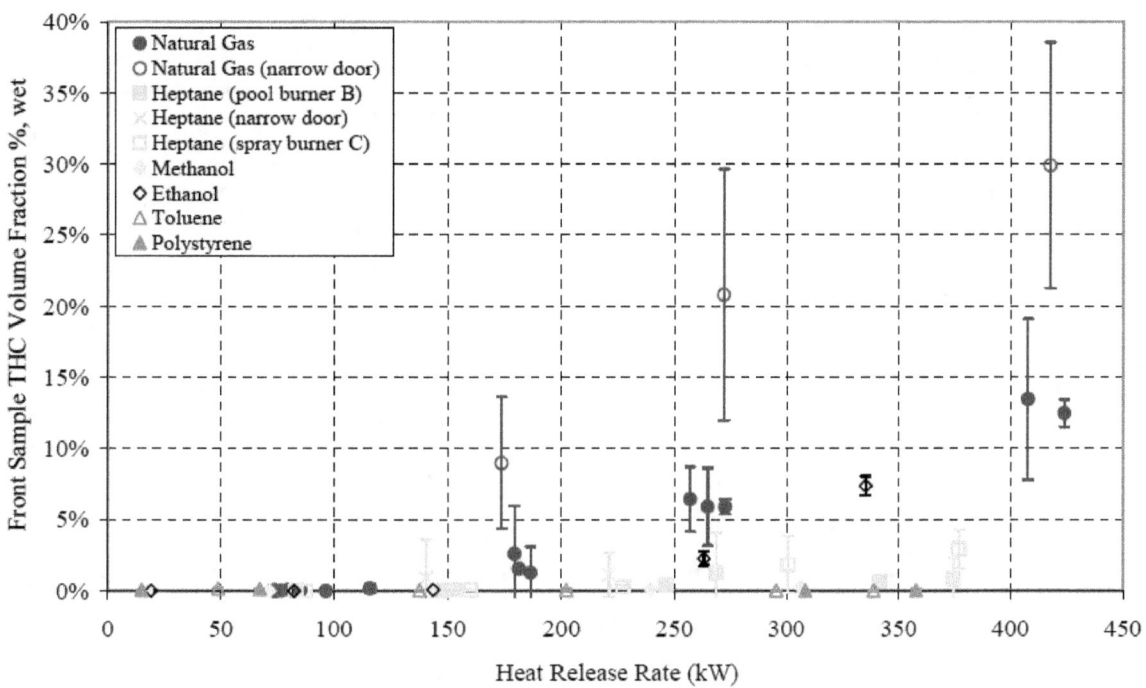

Figure 48. Steady state average total hydrocarbon volume fraction measurements at front sample probe location.

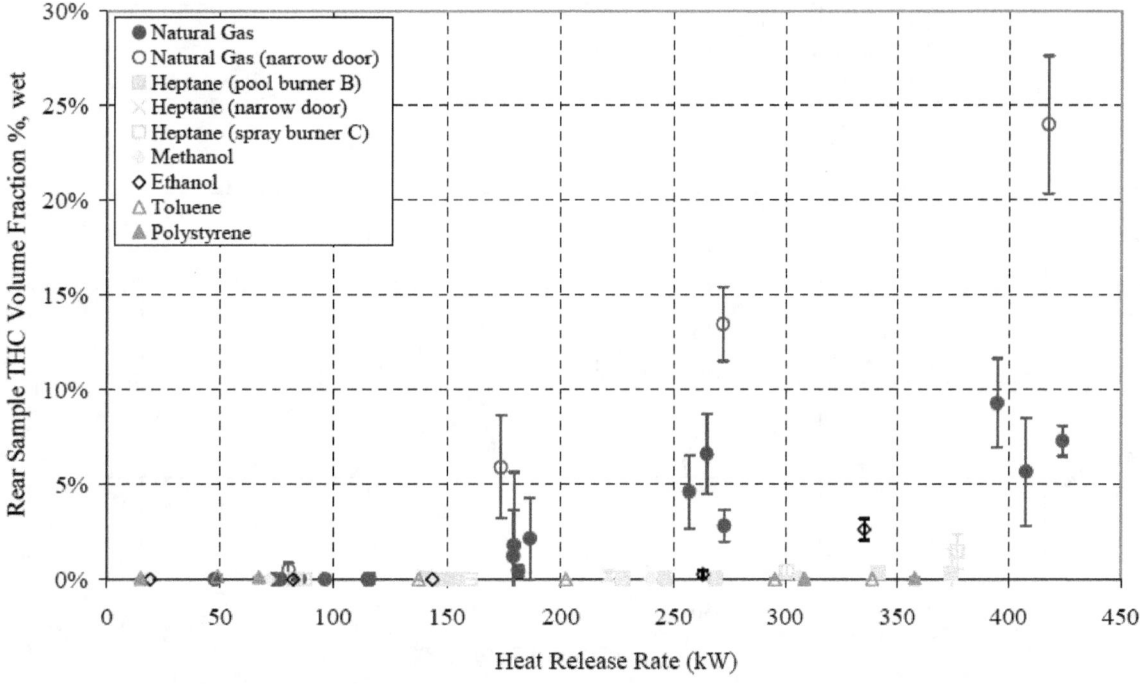

Figure 49. Steady state average total hydrocarbon volume fraction measurements at rear sample probe location.

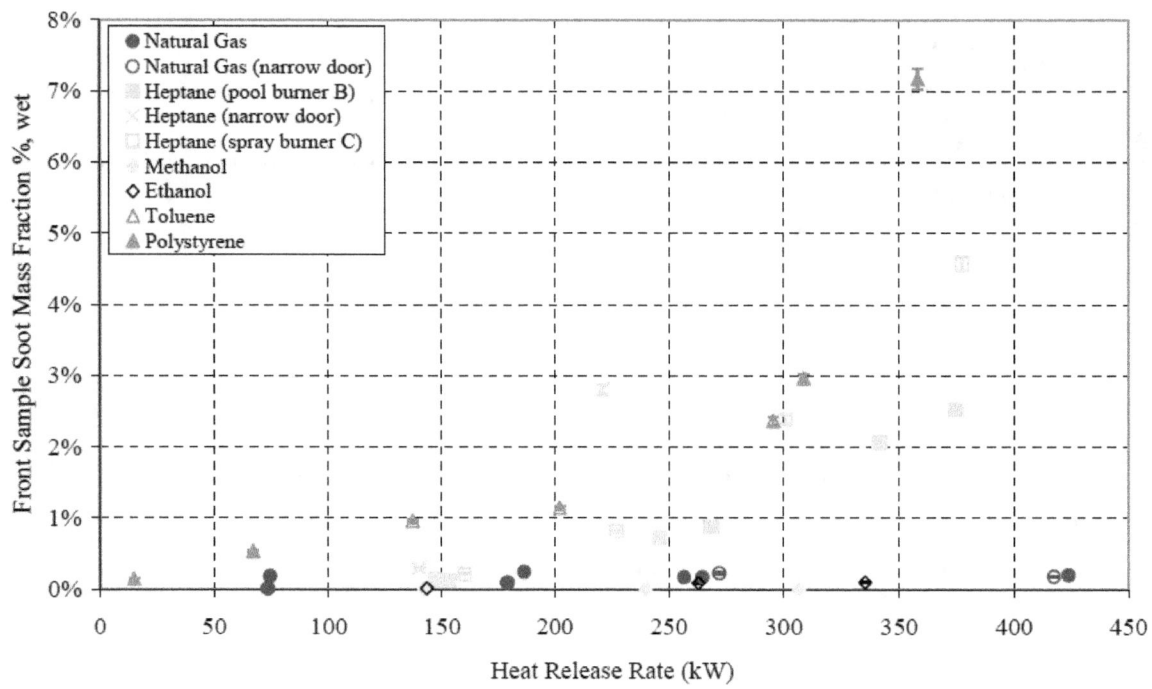

Figure 50. Steady state gravimetric soot mass fraction measurements at front sample probe location.

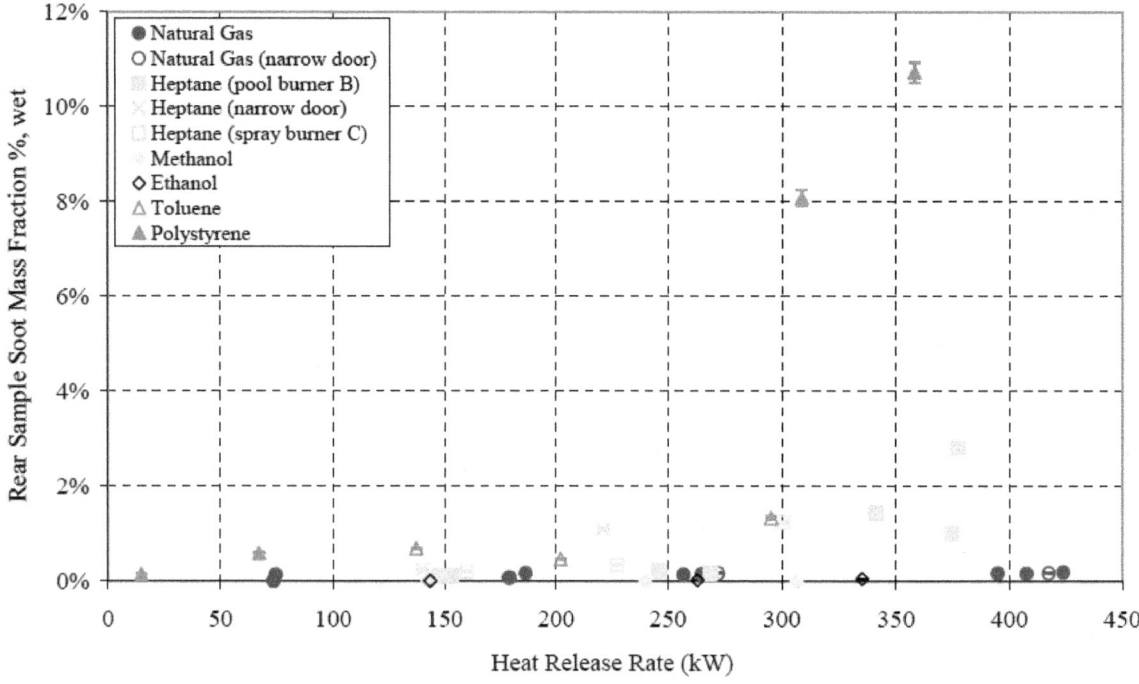

Figure 51. Steady state gravimetric soot mass fraction measurements at rear sample probe location.

Table 15. Summary of steady state rear gas and soot sample probe measurements.

Test #	Fuel	SS Window		HRR (kW)		O₂ Rear (%)		CO₂ Rear (%)		CO Rear (%)		THC Rear (%)		Soot Rear g/g (%)	
		start (s)	stop (s)	Mean	U	Mean	U	Mean	U	Mean	U	Mean	U	Mean	U
1	Natural Gas	500	1924	75	11	9.86	0.77	5.20	0.38	0.11	0.04	0.01	0.01	0.14	0.01
		2220	3384	186	23	0.12	0.12	8.88	0.28	1.22	0.42	2.16	2.14	0.17	0.01
		3600	3944	77	11	10.69	1.37	4.77	0.70	0.06	0.03	0.01	0.01		
2	Natural Gas	400	1594	257	32	0.05	0.12	8.12	0.61	1.63	0.40	4.60	1.94	0.14	0.01
		1840	3229	395	49	0.06	0.15	7.17	0.48	2.47	0.31	9.30	2.33	0.17	0.01
		3500	4454	179	23	0.17	0.52	8.87	0.20	0.75	0.27	1.23	2.42	0.08	0.01
		4550	5004	115	15	4.91	1.06	7.41	0.49	0.02	0.02	0.02	0.04		
		5200	5579	48	7	13.76	0.43	3.34	0.22	0.05	0.03	0.01	0.01		
3	Natural Gas	475	1139	265	33	0.12	0.10	7.93	0.22	1.69	0.21	6.62	2.11	0.15	0.01
		1300	2224	408	55	0.06	0.10	7.06	0.28	2.40	0.22	5.68	2.85	0.17	0.01
		2555	3449	179	25	0.31	0.25	8.79	0.20	0.77	0.25	1.80	3.87	0.08	0.01
		3645	4019	116	14	5.64	1.26	7.07	0.56	0.01	0.03	0.03	0.28		
		4390	5249	74	10	10.99	0.69	4.59	0.32	0.01	0.02	0.00	0.01	0.01	0.01
4	Heptane (pool)	1375	2149	153	21	7.37	1.13	8.28	0.69	0.06	0.04	0.02	0.12	0.11	0.01
		2850	3334	269	34	7.38	2.26	8.52	1.38	0.05	0.07	0.05	0.13	0.17	0.01
		4090	5489	375	54	3.20	1.87	9.80	0.93	0.64	0.51	0.29	1.34	1.01	0.02
5	Heptane (pool)	1245	1799	140	18	3.18	0.90	10.21	0.48	0.24	0.09	0.15	0.24	0.25	0.01
		2340	2969	221	29	0.16	0.57	9.17	0.40	2.11	0.31	0.21	0.29	1.09	0.02
6	Natural Gas	660	1539	74	11	6.44	1.31	6.67	0.60	0.02	0.03	0.02	0.19	0.01	0.01
		2515	4179	174	22	0.16	0.04	8.00	0.44	1.49	0.37	5.92	2.71		
		4425	4944	272	33	0.14	0.04	7.00	0.23	2.13	0.15	13.47	1.96	0.17	0.01
		5090	5724	417	50	0.09	0.05	6.11	0.31	2.35	0.12	23.99	3.63	0.17	0.01
		6090	6549	80	11	6.52	2.16	6.67	0.97	0.00	0.02	0.51	0.36		
6.5	Natural Gas	285	679	97	12	10.01	1.08	5.42	0.61	0.19	0.06	0.01	0.09		
		920	1204	424	51	0.03	0.04	7.62	0.41	2.17	0.14	7.31	0.80	0.19	0.01
		1600	2329	273	33	0.10	0.13	8.33	0.23	1.37	0.24	2.82	0.83		
		2540	2804	181	22	1.99	1.42	8.51	0.52	0.36	0.18	0.46	0.28		
		2980	3259	85	11	11.48	0.78	4.44	0.39	0.07	0.07	0.02	0.01		
7	Heptane (pool)	1200	1669	148	20	9.18	0.67	6.92	0.38	0.03	0.02	0.01	0.08	0.11	0.01
		2105	2664	246	38	8.36	2.24	7.24	1.00	0.14	0.21	0.07	0.10	0.23	0.01
		3040	3709	341	43	3.46	1.94	9.08	0.53	0.99	0.66	0.38	0.27	1.42	0.03
8	Methanol (pool)	1439	2009	17	3	18.97	0.27	1.08	0.10	0.00	0.02	0.00	0.01		
9	Ethanol (pool)	1300	2019	19	3	18.82	0.21	1.23	0.06	0.00	0.02	0.00	0.01		
10	Toluene (pool)	1400	1884	49	9	15.74	0.63	3.58	0.41	0.21	0.09	0.18	0.05		
		2805	3154	138	20	11.24	0.55	6.86	0.46	0.16	0.03	0.01	0.01	0.69	0.01
		3600	4224	202	27	9.30	0.88	8.43	0.67	0.04	0.03	0.01	0.01	0.47	0.01
		4435	5044	295	43	3.74	1.35	12.33	0.86	0.18	0.11	0.01	0.01	1.33	0.03
		5120	5394	339	44	2.19	0.76	12.76	0.39	0.65	0.40	0.01	0.01		
11	Ethanol (spray)	550	1039	83	15	12.57	1.35	4.92	0.79	0.29	0.27	0.01	0.01		
		1400	1714	144	18	8.62	1.19	7.27	0.71	0.02	0.03	0.01	0.13	0.01	0.01
		2175	2849	263	34	1.39	1.34	11.06	0.62	0.55	0.37	0.26	0.21	0.03	0.01
		2940	4200	335	51	0.00	0.03	9.80	0.43	3.60	0.67	2.64	0.57	0.05	0.01
12	Methanol (spray)	300	724	72	10	14.30	0.37	3.63	0.18	0.00	0.02	0.00	0.01		
		1145	1609	143	19	7.61	1.30	7.35	0.70	0.02	0.03	0.00	0.01	0.00	0.01
		1949	2669	240	30	0.07	0.18	9.61	0.76	2.97	1.34	0.05	0.03	0.00	0.01
		2760	3299	306	38	0.00	0.02	6.70	0.38	7.75	0.51	0.18	0.02	0.00	0.01
13	Polystyrene	710	1344	15	3	19.03	0.26	1.34	0.12	0.07	0.02	0.07	0.01	0.14	0.01
14	Polystyrene	870	1724	67	12	15.20	0.72	4.02	0.51	0.21	0.03	0.12	0.01	0.59	0.01
15	Heptane (spray)	280	759	88	16	12.32	0.73	5.07	0.41	0.02	0.02	0.00	0.01		
		950	1259	160	19	9.72	0.87	6.54	0.50	0.04	0.02	0.01	0.01	0.20	0.01
		1475	1999	227	30	7.93	2.75	7.57	1.37	0.06	0.13	0.03	0.08	0.34	0.01
		2200	2764	301	46	0.93	1.22	10.66	0.52	0.76	0.44	0.43	0.35	1.24	0.03
		2790	3169	377	58	0.24	0.49	9.93	0.76	1.70	0.85	1.47	0.92	2.80	0.06
		3390	3734	83	15	12.76	0.66	4.75	0.38	0.04	0.03	0.03	0.01		
16	Polystyrene	545	649	358	42	0.16	0.13	9.36	0.76	3.93	0.64	0.08	0.04	10.72	0 21
		715	769	309	35	0.12	0.03	10.89	0.79	2.81	0.51	0.04	0.01	8.07	0.16

Table 16. Summary of steady state front gas and soot sample measurements.

Test #	Fuel	SS Window		HRR (kW)		O₂ Rear (%)		CO₂ Rear (%)		CO Rear (%)		THC Rear (%)		Soot Rear g/g (%)	
		start (s)	stop (s)	Mean	U	Mean	U	Mean	U	Mean	U	Mean	U	Mean	U
1	Natural Gas	500	1924	75	11	12.45	0.63	3.98	0.29	0.07	0.03	0.02	0 18	0.18	0.01
		2220	3384	186	23	3.14	1.06	7.54	0.41	0.87	0.32	1.27	1.83	0.25	0.01
		3600	3944	77	11	13.07	0.65	3.81	0.30	0.06	0.03	0.03	0 28		
2	Natural Gas	400	1594	257	32	0.20	0.38	7.27	0.35	2.12	0.40	6.44	2.29	0.17	0.01
		1840	3229	395	49	0.22	0.34	5.92	0.33	3.28	0.25	0.00	0.00		
		3500	4454	179	23	3.34	0.86	7.12	0.30	0.85	0.25	0.00	0.00	0.09	0.01
		4550	5004	115	15	8.12	0.74	5.66	0.32	0.14	0.07	0.00	0.00		
		5200	5579	48	7	15.01	0.32	2.68	0.14	0.04	0.02	0.00	0.00		
3	Natural Gas	475	1139	265	33	0.60	0.35	7.28	0.22	1.97	0.27	5.91	2.72	0.17	0.01
		1300	2224	408	55	0.12	0.10	6.08	0.33	3.19	0.20	13.45	5.67	0.00	0.00
		2555	3449	179	25	2.87	0.65	7.42	0.17	0.90	0.21	2.59	3.37	0.09	0.01
		3645	4019	116	14	7.98	0.58	5.84	0.23	0.13	0.05	0.18	0.33		
		4390	5249	74	10	12.92	0.49	3.70	0.21	0.02	0.02	0.03	0 39	0.01	0.01
4	Heptane (pool)	1375	2149	153	21	8.21	0.97	7.63	0.56	0.05	0.03	0.07	0 14	0.12	0.01
		2850	3334	269	34	0.83	1.86	11 12	1.22	1.63	1.16	1.26	2.85	0.87	0.02
		4090	5489	375	54	0.23	0.34	9.94	0.61	1.89	0.58	0.82	1.46	2.52	0.05
5	Heptane (pool)	1245	1799	140	18	2.30	0.78	10 94	0.44	0.15	0.09	0.92	2.72	0.29	0.01
		2340	2969	221	29	1.56	1.91	10 37	1.14	0.87	0.69	1.13	1.54	2.80	0.07
6	Natural Gas	660	1539	74	11	8.64	0.76	5.66	0.34	0.02	0.03	0.04	0.27	0.01	0.01
		2515	4179	174	22	0.36	0.25	7.73	0.44	1.65	0.46	8.99	4.63		
		4425	4944	272	33	0.10	0.13	6.81	0.57	2.31	0.39	20.79	8.83	0.23	0.01
		5090	5724	417	50	0.04	0.04	6.34	0.41	2.45	0.18	29.89	8.65	0.18	0.01
		6090	6549	80	11	7.92	0.72	5.99	0.32	0.02	0.04	0.10	0.08		
6.5	Natural Gas	285	679	97	12	11.13	0.62	4.71	0.31	0.10	0.03	0.01	0.01		
		920	1204	424	51	0.13	0.10	5.97	0.33	2.94	0.20	12.47	0 95	0.20	0.01
		1600	2329	273	33	0.51	0.15	6.80	0.14	2.41	0.13	5.90	0.51		
		2540	2804	181	22	2.59	0.16	7.39	0.11	1.10	0.09	1.55	0.25		
		2980	3259	85	11	12.29	0.64	4.02	0.29	0.03	0.03	0.05	0.02		
7	Heptane (pool)	1200	1669	148	20	9.10	0.59	6.92	0.33	0.05	0.03	0.03	0.02	0.13	0.01
		2105	2664	246	38	3.09	1.08	9.68	0.33	0.65	0.27	0.40	0.16	0.72	0.02
		3040	3709	341	43	0.71	0.36	9.45	0.37	1.97	0.39	0.64	0.20	2.06	0.04
8	Methanol (pool)	1439	2009	17	3	19.15	0.27	1.01	0.10	0.00	0.02	0.00	0.01		
9	Ethanol (pool)	1300	2019	19	3	19.02	0.21	1.15	0.06	0.00	0.02	0.00	0.01		
10	Toluene (pool)	1400	1884	49	9	16.46	0.62	3.08	0.43	0.20	0.10	0 17	0.03		
		2805	3154	138	20	8.06	0.80	9.05	0.39	0.20	0.03	0.05	0.01	0.96	0.02
		3600	4224	202	27	3.55	0.87	12.32	0.60	0.15	0.05	0.06	0.03	1.14	0.02
		4435	5044	295	43	1.21	0.55	13.43	0.39	0.52	0.13	0.03	0.02	2.37	0.05
		5120	5394	339	44	0.28	0.38	13.10	0.47	1.36	0.49	0.01	0.01		
11	Ethanol (spray)	550	1039	83	15	14.05	0.69	4.03	0.39	0.12	0.09	0.01	0.01		
		1400	1714	144	18	8.07	0.86	7.42	0.40	0.20	0.15	0.08	0.08	0.01	0.01
		2175	2849	263	34	0.98	0.44	9.05	0.30	3.64	0.64	2.27	0.50	0.09	0.01
		2940	4200	335	51	0.05	0.08	6.86	0.28	7.60	0.47	7.39	0.67	0.10	0.01
12	Methanol (spray)	300	724	72	10	15.18	0.36	3.17	0.17	0.01	0.02	0.00	0.01		
		1145	1609	143	19	7.70	0.96	7.01	0.38	0.40	0.23	0.01	0.01	0.00	0.01
		1949	2669	240	30	1.11	0.69	7.68	0.57	4.83	1.39	0.09	0.03	0.00	0.01
		2760	3299	306	38	0.25	0.29	5.75	0.96	8.52	1.96	0.23	0.08	0.00	0.01
13	Polystyrene	710	1344	15	3	19.22	0.25	1.24	0.11	0.07	0.02	0.07	0.01	0.15	0.01
14	Polystyrene	870	1724	67	12	15.66	0.60	3.70	0.39	0.19	0.03	0 14	0.01	0.53	0.01
15	Heptane (spray)	280	759	88	16	13.32	0.55	4.47	0.31	0.01	0.02	0.01	0.01		
		950	1259	160	19	7.65	0.48	7.69	0.28	0.08	0.03	0.06	0.04	0.21	0.01
		1475	1999	227	30	3.46	1.05	9.58	0.46	0.45	0.21	0.26	0.15	0.82	0.02
		2200	2764	301	46	0.32	0.19	8.42	1.25	2.82	1.01	1.83	2.00	2.38	0.05
		2790	3169	377	58	0.02	0.08	7.00	1.08	3.96	0.80	2.91	1.35	4.58	0.09
		3390	3734	83	15	13.38	0.51	4.23	0.23	0.05	0.05	0.04	0.03		
16	Polystyrene	545	649	358	42	0.14	0.10	10.32	0.54	3.13	0.39	0.05	0.03	7.17	0.14
		715	769	309	35	0.26	0.13	12.24	1.26	1.76	0.89	0.03	0.02	2.96	0.06

60

3.4.1 Gas Chromatography

The results of the gas chromatography (GC) analysis are given in Table 17 below. All gas volume fraction values listed in this table and in the following figures are on a wet basis. The measurements listed in the table were multiplied by a factor of 10^6. The GC was connected to the front probe gas sample line using the outlet of the bypass line on the total hydrocarbon analyzer. The front probe location is given in Table 2 and shown in Fig. 4.

The injection times were manually recorded relative to the fire ignition time. Because the turnaround time for the GC sampling was greater than 20 min, a limited number of samples were acquired. For comparison with the GC results, the HRR, the total hydrocarbons (FID), and the carbon monoxide volume fraction were averaged over a 30 s window ending at the injection time. The total hydrocarbon volume fractions determined using the GC ("Total HC, from GC" in Table 17) were calculated for comparison with the total hydrocarbon analyzer results. This value was determined by summing each of the measured hydrocarbon volume fractions after converting to an equivalent methane basis. The results of this comparison are illustrated in Fig. 52.

For most of the GC samples acquired, the two methods agreed within experimental uncertainty. This confirms that the compounds occurring in significant amounts were correctly accounted for in the GC measurement. The most notable exception to this is the GC sample acquired during the natural gas test #3 at t = 2140 s, where the results differed by a factor of 5 for unknown reasons. For the polystyrene tests, the total hydrocarbons measurements from the GC were significantly lower than from the total HC analyzer. Since the GC method did not include compounds above C_6, the presence of very large hydrocarbons could explain the differences, although this is an unlikely explanation. Future work will revisit this issue. The values in Table 17 that are in bold type are below the quantifiable detection limits of the method (the uncertainty was greater than the measured value), but were included to show that they were identified in trace amounts. The results of the uncertainty analysis of the GC measurements (described in Sec. 2.5.7) are show in Table 18. The values represent the combined expanded uncertainty with a coverage factor of 2.

The GC results for the fire tests using natural gas showed, to a large degree, that the hydrocarbons measured at the sample location were simply unburned fuel. Figure 53 shows the composition of hydrocarbon species for a typical natural gas test. For the liquid and solid fuels included in this study, a large number of intermediate hydrocarbon species were quantified. In all cases, methane was the largest measured component of hydrocarbon species. Even though the GC measured species only as large as C_6, the similarity between the total hydrocarbon results from the GC and the total hydrocarbon analyzer (see Fig. 52) provides evidence that there were no species of significant quantity missed by the GC analysis. These results imply that in the upper layer of these compartment fires methane was the most abundant hydrocarbon species, higher in concentration than the parent fuel in all cases. For example, Fig. 54 shows the GC measurements of the most abundant species (methane, ethyne plus ethene, and benzene) for several fire sizes, burning heptane as the fuel. Other species were below the GC detection limits (see blank spaces in Table 17). For the fires burning the two alcohol fuels, there is some uncertainty with regard to the total hydrocarbon and GC results, since the Nafion filter is known to absorb polar organic compounds such as alcohols.

61

Table 17. Summary of GC sample results. Values in bold were identified as trace species. Blank spaces imply that the measurements were below the detection limits.

Test #	Fuel	GC Injection Time (s)	HRR (kW)	Total HC, FID, (mol/mol)*1e6	Total HC, from GC, (mol/mol)*1e6	Carbon Monoxide, (mol/mol)*1e6	Methane, (mol/mol)*1e6	Ethyne + Ethene, (mol/mol)*1e6	Ethane, (mol/mol)*1e6	Propene, (mol/mol)*1e6	Propane, (mol/mol)*1e6	Propyne, (mol/mol)*1e6	1,3-butadiene, (mol/mol)*1e6	n-Pentane, (mol/mol)*1e6	Benzene, (mol/mol)*1e6
1	Natural Gas	2940	184.3	11881	16590	7785	12058	397	1456						138
2	Natural Gas	970	255.8		21089	19687	16993	423	1164						154
		3130	391.4		51188	32685	42152	3330	**58**	37		62	**11**		320
		4330	175.7		8642	9203	6685	734	**29**			**12**			**66**
3	Natural Gas	910	272.1	45673	30372	19356	23849	2432	83	19		35	**4**		219
		2140	405.5	116515	26511	32587	21920	1733	**50**	21		32	**6**		140
		3375	176.3	29487	11240	10055	8301	1121	**35**			18	**8**		91
6.5	Natural Gas	1065	429.1	131001	137260	29794	134908	11770	905	216	**51**	365	64	**10**	752
		2310	272.5	60342	56684	24537	52668	6373	138	50		184	**20**		496
7	Heptane	1150	145.6	300	155	396	-2	70							5
		2460	222.9	2967	2599	4363	793	756				5	**3**		100
		3675	341.5	5563	5064	20178	2944	1165				6			89
10	Toluene	2840	131.5	580	469	2015	**8**	30							71
		4055	199.5	488	410	1257	**53**	34							53
11	Ethanol	1500	142.2	601	825	2054	270	252							9
		2740	268.5	20298	20618	33146	8332	5378	**19**	11		48	**5**		216
		3950	348.0	72784	80041	74807	33342	20725	156	76		134	58	**10**	679
12	Methanol	2025	236.8	870	805	47196	706	50							
		3255	309.7	1923	2066	77652	1801	132							
13	Polystyrene	1110	14.8	745	74	686	-23	48							
14	Polystyrene	1155	63.2	1351	748	1777	-4	114					**1**		86
15	Heptane	1565	228.1	4236	4021	7004	1744	995				7			134
		3090	396.6	31937	26328	43786	13656	4954			**6**	43	**12**		1044
16	Polystyrene	620	355.4	502	309	32227	268	21							

note: negative values listed in this table are a result of measurement uncertainty associated with the baseline drift correction of the quantification method and are not physically meaningful.

Table 18. Summary of GC measurement uncertainty analysis results. Blank spaces imply that the measurements were below the detection limits.

Test #	Fuel	GC Injection Time (s)	Methane, (mol/mol)*1e6	Ethyne + Ethene, (mol/mol)*1e6	Ethane, (mol/mol)*1e6	Propene, (mol/mol)*1e6	Propane, (mol/mol)*1e6	Propyne, (mol/mol)*1e6	1,3-butadiene, (mol/mol)*1e6	n-Pentane, (mol/mol)*1e6	Benzene, (mol/mol)*1e6
1	Natural Gas	2940	123	16	64						73
2	Natural Gas	970	152	16	57						72
		3130	342	40	67	14		14	36		67
		4330	103	15	68			15			76
3	Natural Gas	910	199	28	66	15		14	36		70
		2140	185	20	67	15		14	36		73
		3375	108	15	67			15	36		75
6.5	Natural Gas	1065	1114	162	53	13	65	12	35	75	61
		2310	427	83	64	14		13	36		63
7	Heptane	1150	104	18							78
		2460	102	14				15	36		74
		3675	100	15				15			75
10	Toluene	2840	104	18							75
		4055	104	18							76
11	Ethanol	1500	103	17							78
		2740	108	69	68		15	14	36		70
		3950	271	293	64		14	14	35	75	61
12	Methanol	2025	102	18							
		3255	100	18							
13	Polystyrene	1110	104	18							
14	Polystyrene	1155	104	18					36		75
15	Heptane	1565	101	15				15			73
		3090	132	63	68	15		14	36		62
16	Polystyrene	620	103	18							

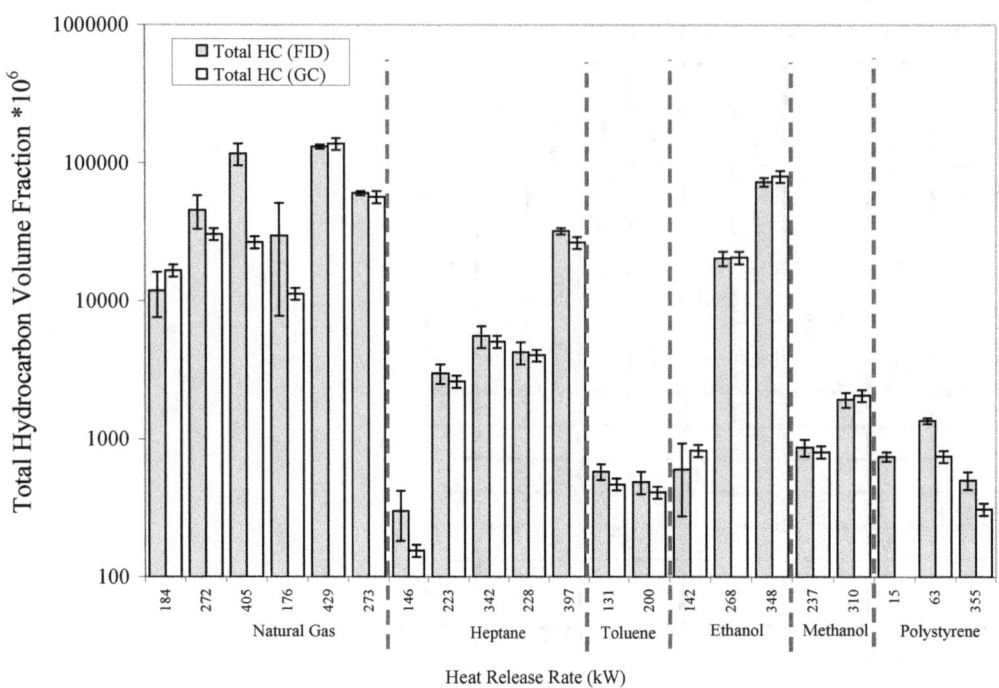

Figure 52. Comparison of total hydrocarbons measured using the GC and the total hydrocarbon analyzer (THC Front), both expressed on a CH_4 basis.

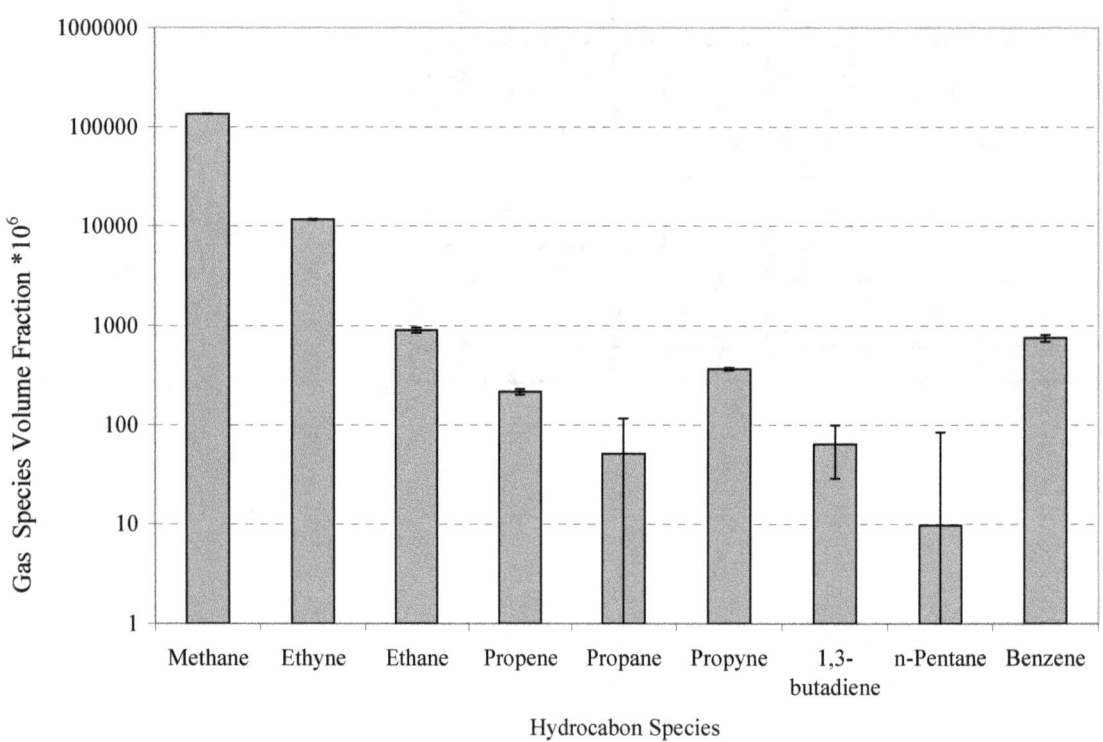

Figure 53. GC gas species composition measurements in the natural gas fire #65, HRR=429 kW.

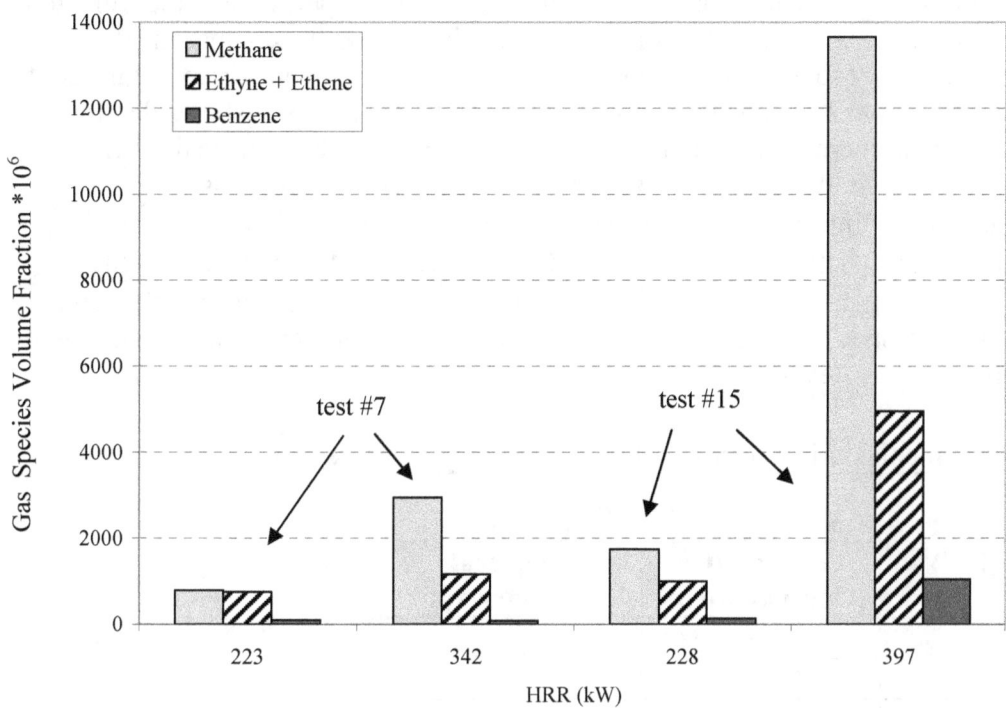

Figure 54 GC measurements of methane, ethyne plus ethene, and benzene in the heptane pool fire (test #7) and heptane spray fire (test #15), front gas sample probe.

3.5 Optical Soot

Figure 55 shows the time variation of the optical measurement of soot mass concentration for test #6, a natural gas fire with a nominal heat release rate that varied from 75 kW to 400 kW. Table 19 presents the average optically measured soot mass concentrations in the doorway for the four fire sizes considered in test #6. The optical measurement averages were determined during the same time periods as the gravimetric measurements. Figure 56 shows the measured soot concentration varied from 47 mg/m^3 to 380 mg/m^3, following the trend in fire size.

Figure 57 shows the variation with time of the soot mass concentration for the 20 cm diameter polystyrene fire (test #13). For the first 6 min after ignition, the soot mass concentration linearly increased as the fire size increased until a steady state was reached. At 500 s after ignition, the soot mass concentration was almost constant, with an average value of (700 ± 90) mg/m^3 (between 600 s and 1200 s after ignition). As expected, the polystyrene fire was highly sooting, with the soot concentration significantly larger than that of natural gas.

Figure 58 shows the time variation of the soot mass concentration for the 40 cm diameter polystyrene fire (test #14). The mass concentration of soot increased as the fire HRR increased. After ten minutes, the mass concentration of soot stopped increasing, and a near-steady value was reached. The soot mass concentration was approximately (1500 ± 400) mg/m^3 (between 1115 s and 1230 s after ignition). As expected, the doorway of the 40 cm polystyrene pan fire (test #14) had significantly higher soot concentrations than that of the natural gas fire (tests #6) and the 22 cm polystyrene pan fire (test #13).

65

Table 19 also shows the gravimetric measurements made inside the compartment at the front and rear locations. Both sets of data show that as the fire became larger and the compartment became underventilated, higher soot mass concentrations were present. A comparison of the soot mass concentrations at the front and rear of the compartment with the doorway shows that the amount inside of the compartment was larger by a factor of 2 to 3 for all but the smallest fire sizes. As was true for the larger natural gas fires in test #6, the observed soot mass concentrations inside of the compartment in tests #13 and 14 (1140 mg/m^3 ± 140 mg/m^3 and 2700 mg/m^3 ± 140 mg/m^3, respectively) were much (a factor of 1.6 to 1.8) larger than in the doorway (700 mg/m^3 ± 90 mg/m^3 and 1500 mg/m^3 ± 400 mg/m^3, respectively) during the same sampling periods in these tests. The larger concentrations within the compartment may be due to air entrainment in the doorway, leading to soot oxidation.

Table 19. Soot measurements during narrow doorway natural gas fire (test #6).

Gravimetric Sampling Location	HRR (kW)	Soot (mg/m^3)	
		Gravimetric (Compartment)	Optical (Doorway)
rear	75	35 ± 62 %	47 ± 13 %
front		31 ± 62 %	
rear	180	428 ± 7 %	147 ± 17 %
front		390 ± 7 %	
rear	270	763 ± 7 %	270 ± 18 %
front		702 ± 7 %	
rear	400	766 ± 7 %	380 ± 17 %
front		582 ± 7 %	

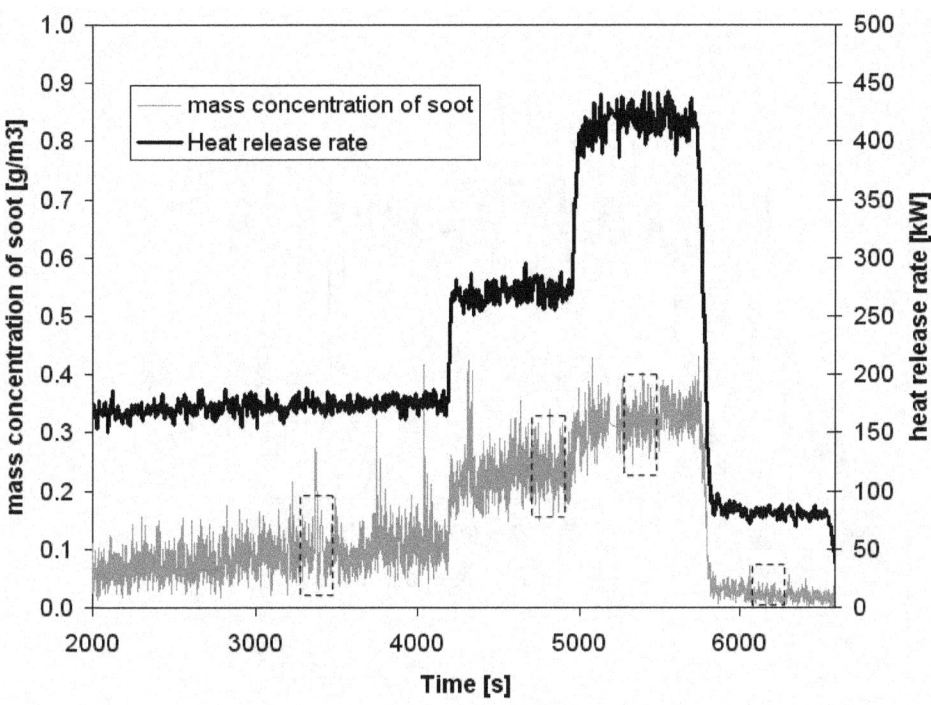

Figure 55. History of the soot mass concentration during natural gas fire (test #6). Mean values of the soot mass concentration were determined over the same periods as the gravimetric measurements as indicated by the dotted boxes. The fire heat release rate is also shown.

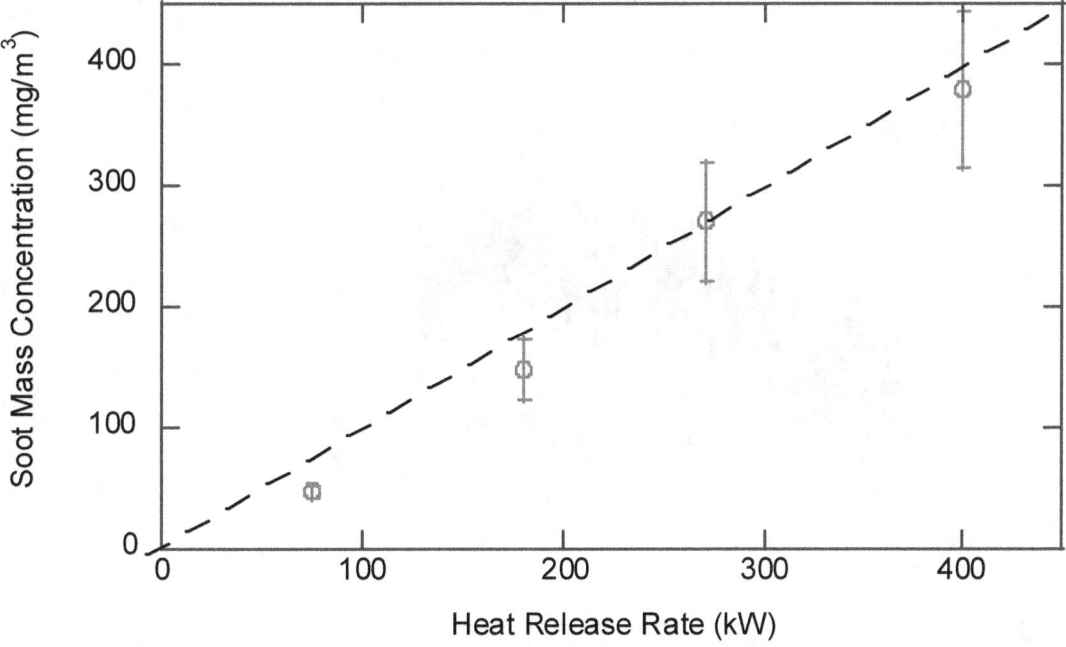

Figure 56. Soot mass concentration in the doorway during natural gas fire (test #6) as a function of the HRR.

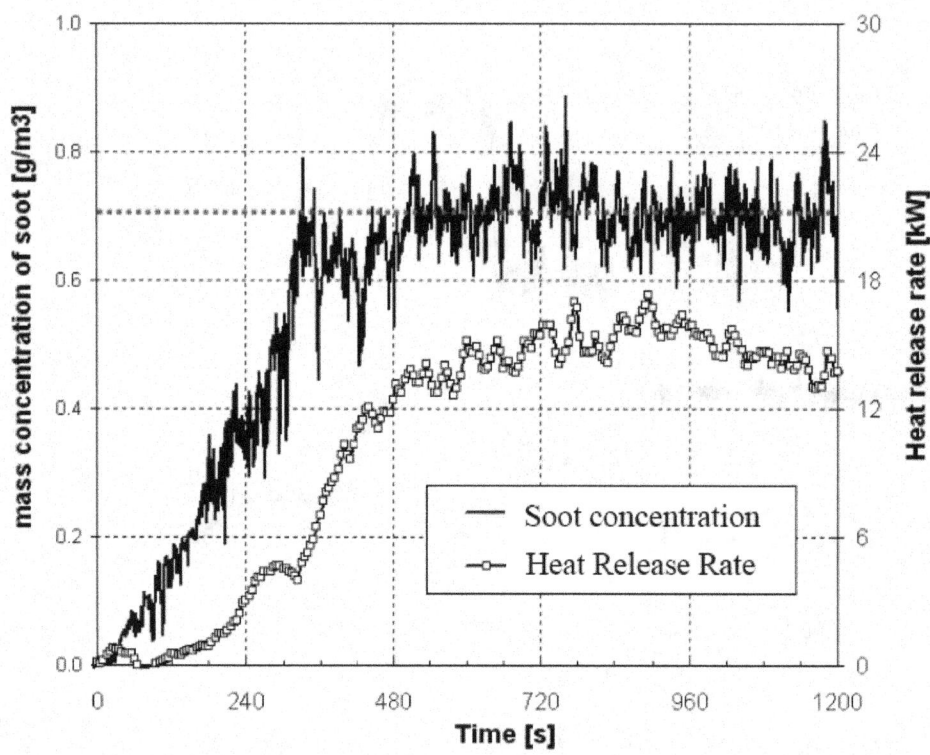

Figure 57. History of the particle mass fraction in the doorway during test #13, polystyrene burning in the 20 cm diameter burner. The fire heat release rate is also shown.

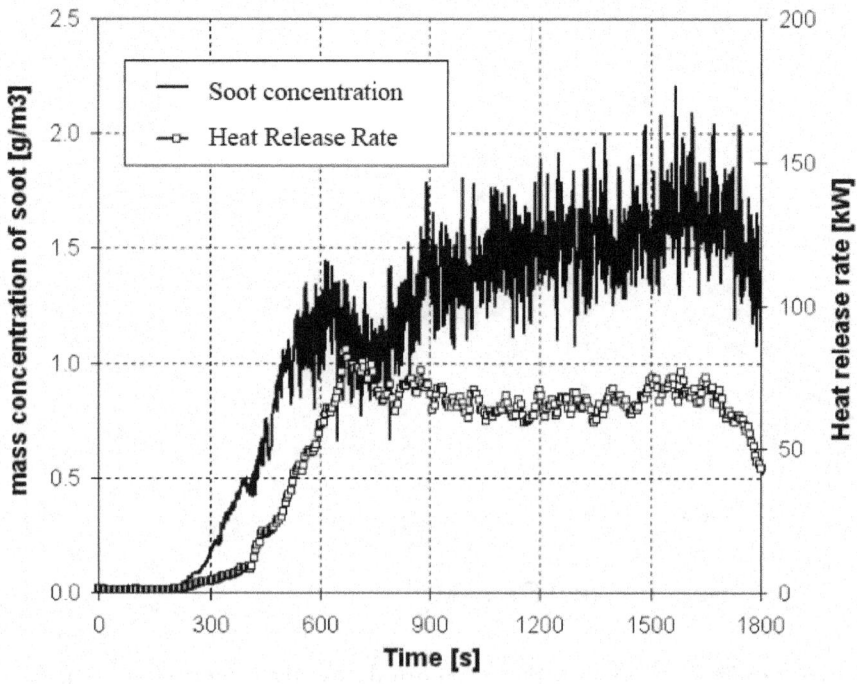

Figure 58. History of the soot mass concentration in the doorway during test #14, polystyrene burning in a 40 cm diameter burner. The fire heat release rate is also shown.

3.6 Heat Fluxes

The heat flux measurements to the floor of the enclosure help to characterize the thermal environment within the enclosure and the transient nature of the interior burning. In an enclosure with a distributed fuel source the heat flux to the floor can be used to predict the onset of flashover. The total heat flux measurement labels are described in Table 20 and their locations are listed in Table 2 of Sec. 2.4. Figure 59 shows the front and rear heat flux results for test #15 with heptane fuel. The front and rear steady state average floor heat flux values for all of the tests are summarized in Table 12 and Table 13 and shown graphically in Fig. 60 and Fig. 61. In general the heat flux levels were significantly higher for the fuels with high soot yields. In addition, for the clean burning fuels the heat flux was fairly constant above heat release rates of 200 kW. Heat flux levels in excess of 200 kW/m^2 were measured for heptane and toluene fires. Although these value are possible based on the measured temperature of the upper layer, they are well beyond the calibrated range of the transducer. Furthermore, all of the heat flux measurements are somewhat artificial since the gauge temperatures were held constant while the floor temperatures were observed to increase significantly. The actual net heat flux to the floor (that has been heated by the fire) would be significantly less than the measured heat flux to the water-cooled gauge.

Table 20. Description of interior total heat flux measurement labels.

Measurement Label	Description
12 HFR (kW/m2)	Total heat flux to floor at rear of enclosure
13 HFF (kW/m2)	Total heat flux to floor at front of enclosure

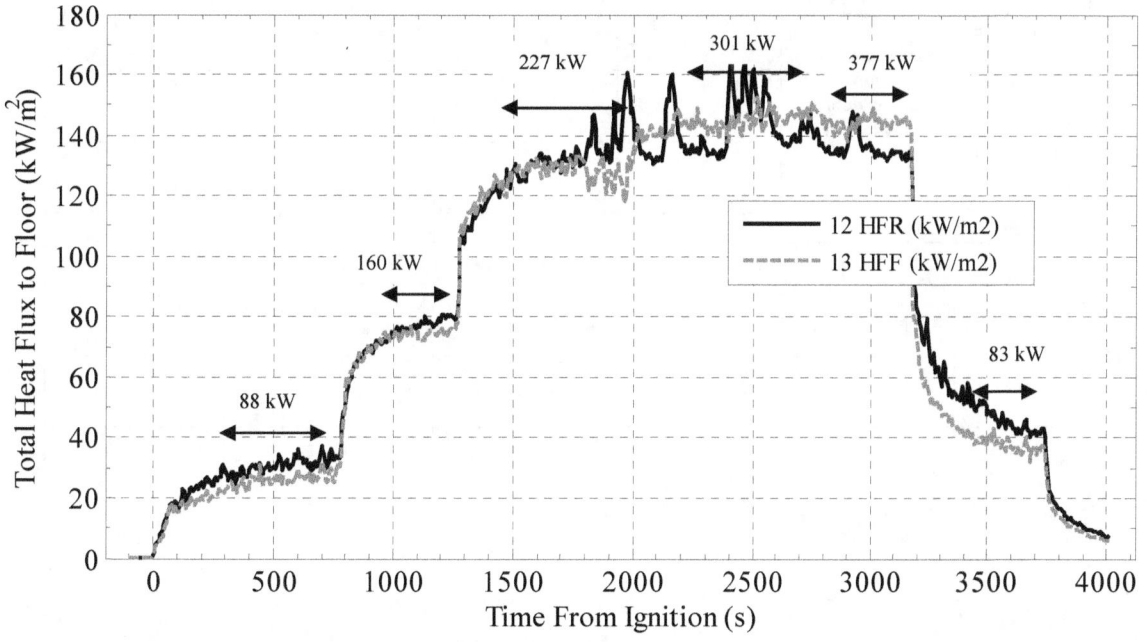

Figure 59. Total heat flux gauge measurements at two locations on the interior floor of the RSE for test #15 using the spray burner with heptane fuel.

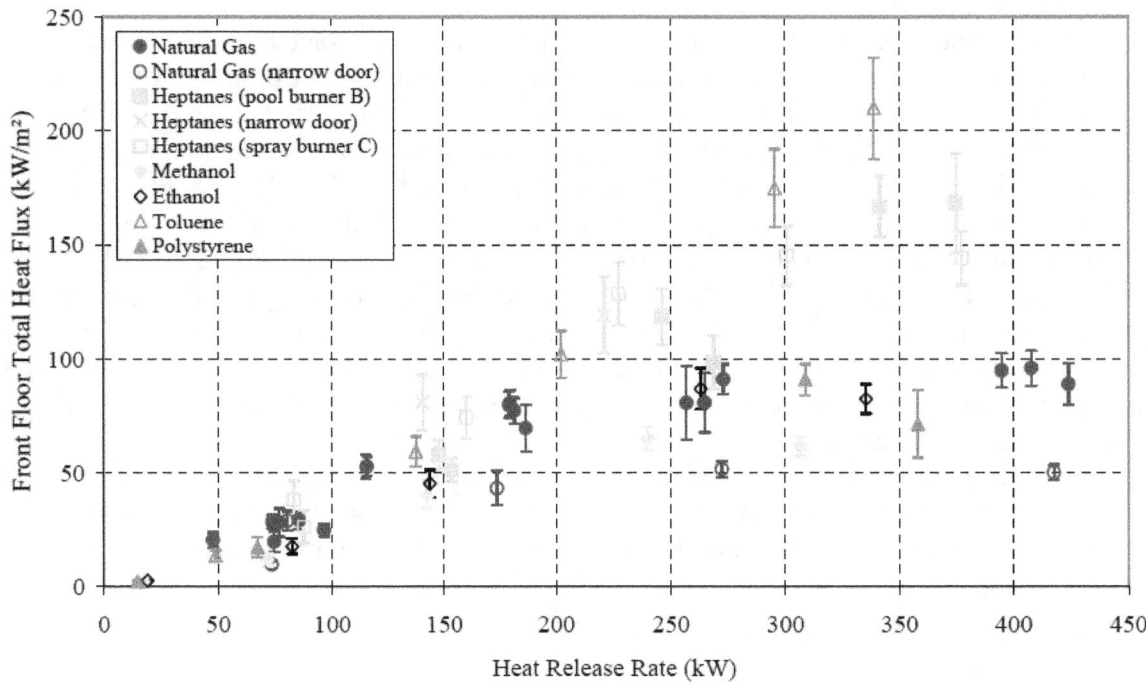

Figure 60 Steady state total heat flux measurement at front floor location. **note:** front heat flux gauge was partially blocked by debris for polystyrene test #16.

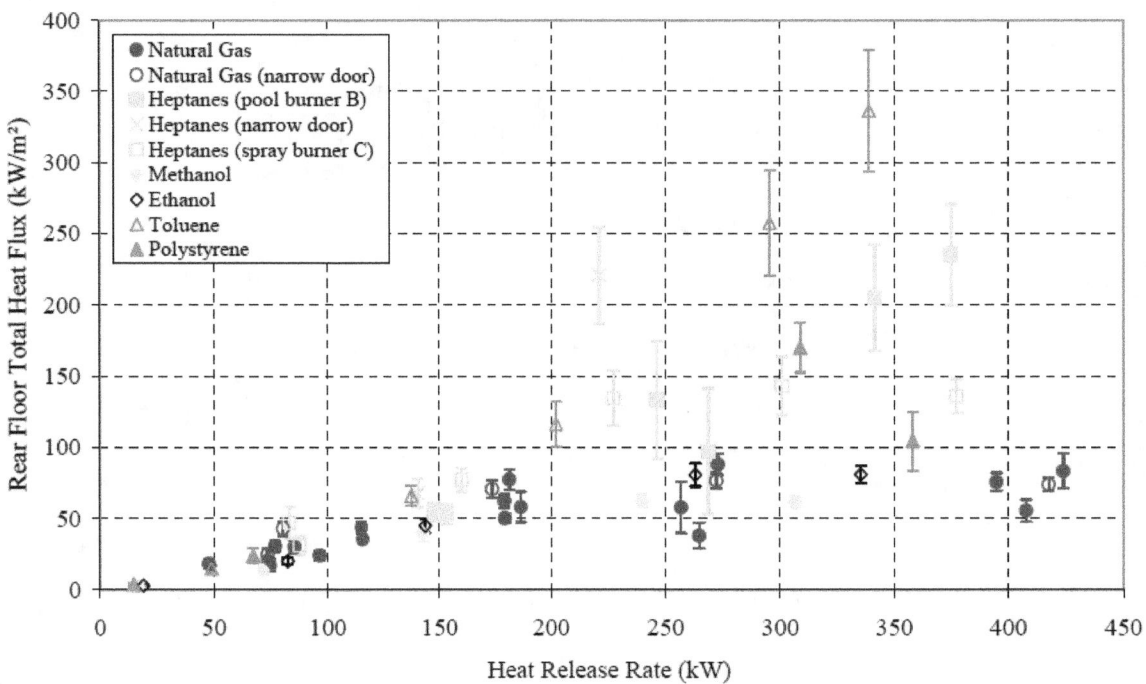

Figure 61. Steady state total heat flux measurement at rear floor location.

4 Discussion of Results

4.1 Fuel and Fire Size Comparisons

Because of the precise metering used, natural gas was the most well controlled fuel used in this study. For these reasons, natural gas was used to observe the effects of changes in burner area, wall material construction, and doorway ventilation. Natural gas produced the highest quantities of total hydrocarbons in this test series. The transition to underventilated burning occurred at a lower HRR for natural gas than for other fuels in this study. Depletion of O_2 and increased CO in the upper layer and flames exiting the doorway are indicators of underventilated burning. The reason for this transition occurring at a lower HRR for natural gas can be explained by differences in fuel stoichiometry (natural gas fuel requires 1 % to 6 % more oxygen for each unit of energy released than the condensed fuels studied) and differences in compartment fire structure.

Like natural gas, liquid heptane fuel was used to study the effects of different burners, door sizes and wall material configurations. Heptane (C_7H_{16}) fuel (a blend of heptane isomers) was selected to represent a moderately sooting liquid alkane. The transition to underventilated burning occurred at a greater HRR for heptane than natural gas. Once the enclosure was underventilated, the CO and THC volume fractions were consistently lower for heptane than natural gas, but the soot concentrations were considerably higher. In general, the gas temperatures and floor heat fluxes were greater and more uniform from front to back for the heptane fires compared to natural gas fires.

Liquid toluene (C_7H_8 - an aromatic hydrocarbon) was included to represent a fuel with a very high soot yield. The toluene fire produced very high temperatures, heat fluxes, and soot concentrations. Temperatures greater than 1500 K were measured, and some of the stainless steel wall pins and aspirated thermocouple probes were melted and destroyed during these tests. Test #10 with toluene appeared to be slightly over-ventilated at the largest fire size, made evident by the presence of oxygen at the rear sample location. However, at the front sample location, oxygen was almost completely depleted and flames were observed outside the doorway. The toluene fires yielded relatively low THC and CO volume fractions compared to the other fuels. One feature that was unique to the toluene test was the formation of a large soot agglomerate (≈10 cm) on the water cooling inlet tube adjacent to the burner (see Fig. 62).

Liquid methanol (CH_3OH) and liquid ethanol (C_2H_5OH) were chosen as clean burning fuels with high combustion efficiency, very low soot yields, and relatively low heats of combustion. The temperatures measured inside the enclosure during the tests with alcohol fuels were similar in magnitude to the tests with natural gas. As expected, there was no measurable amount of soot produced by the methanol fire and the ethanol fire produced soot concentrations similar to the natural gas fires. Extremely high volume fractions of CO were observed for the alcohol fuels after the fires reached ventilation-limited conditions. CO volume fractions greater than 8 % were measured in the front of the enclosure for both of these fuels.

71

Figure 62. Photograph of soot agglomerate on the water inlet tube of the burner after toluene pool fire (test #10).

Polystyrene pellets were burned in a 60 cm diameter pan (test #16). This test differed from all of the other tests, as it involved a solid material in which natural feedback from the fire controlled the mass burning rate. A spray of heptane was ignited and was used to initiate burning of the polystyrene pellets.

The character of the measurements during Test 16 are highlighted in Fig. 63 which shows the measured CO volume fraction in the exhaust stack and at the front and rear of the upper layer in the compartment as a function of time. The data have been time shifted to account for transport delays in the exhaust and sample flows. The figure includes photographs at various times during the experiment which show the appearance of the fire as seen through the open doorway. At both sample locations within the compartment, the CO volume fractions increased as a function of time, reached a maximum, and then decreased to near-zero. In the stack, the CO volume fraction measurements show that the peak at 600 s which was observed inside of the compartment was also reflected in the stack. There were two additional peaks in the stack that were not as pronounced within the compartment.

The photo evidence provides some insight into the CO measurements in the compartment. At 380 s after ignition, the fire was rather small and flames were restricted to locations immediately above the pan. This began to change by 430 s, when the glow of hot gases can be seen in the upper reaches of the compartment until at 600 s, the entire doorway appeared to be filled with flame and smoke rolled out from the top of the doorway. While the doorway was still luminous at 825 s, there were few flames rolling from the doorway. Instead, there was a general glow that was observed. This suddenly changed and by 866 s, the fire had become rather small and the flames were observed to exist only immediately over the pan. At the same time, a steady flow of smoke was transported out from the top part of the compartment.

Observation of the video record does not provide insight into the various extra peaks in the CO measurements in the stack. The reasons for the extra peaks are unclear, but the CO measurements are not an anomaly since the measured soot mass concentration in the exhaust stack tracks the CO peaks observed in the stack. Interestingly, the total hydrocarbon volume fraction has a large peak at 350 s that precedes the first large CO and soot peaks at 380 s. The simultaneous occurrence of the CO and soot peaks is not unexpected. The timing of these concentration peaks are likely related to the compartment fire dynamics. Additional experiments are being planned to investigate this further.

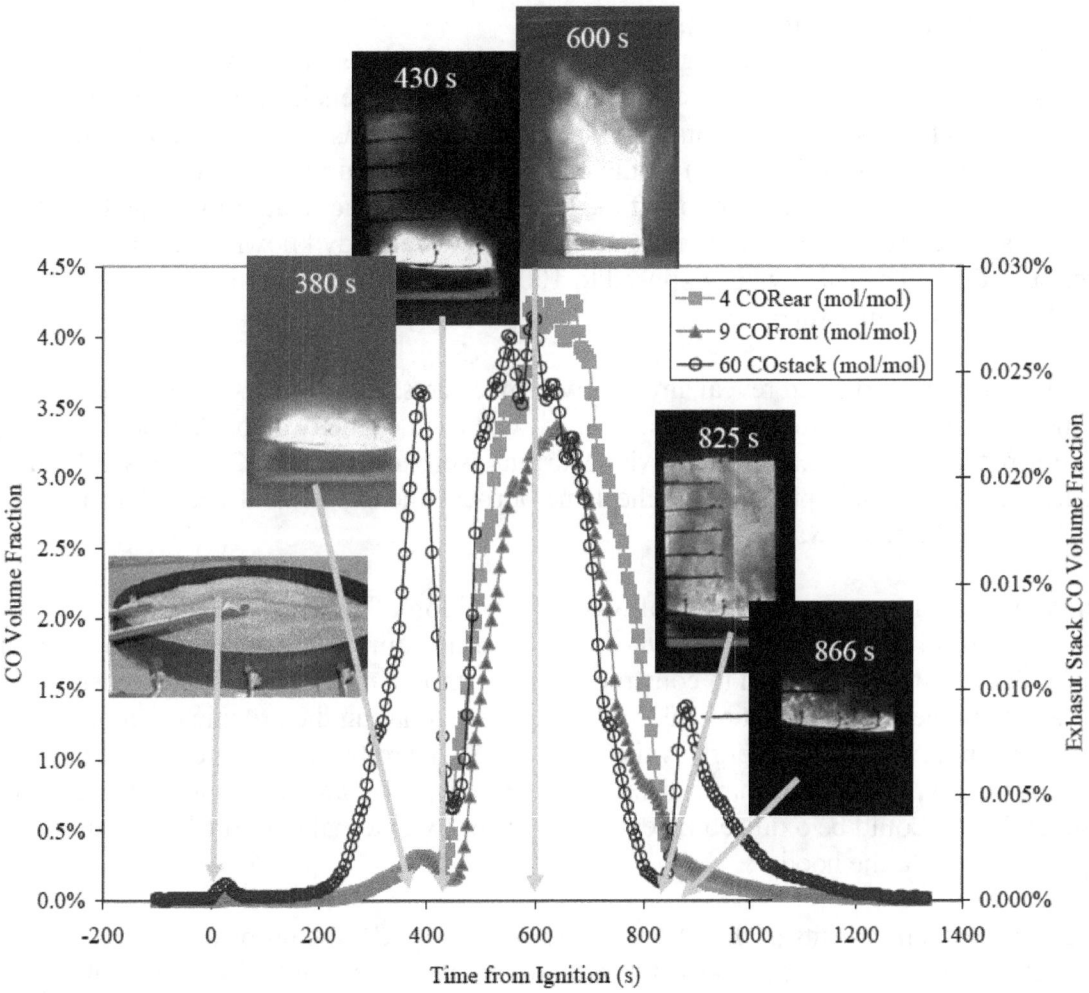

Figure 63 The CO volume fraction measured in the stack and at the front and rear of the compartment as a function of time during the burning of 6.0 kg of polystyrene pellets (Test 16). Photographs at various times show the fire appearance through the open doorway.

4.2 Doorway Ventilation Comparisons

Most of the tests conducted during this series used a 48 cm wide by 81 cm high door vent. Test #5 (heptane) and test #6 (natural gas) were performed using a 24 cm wide by 81 cm high door vent. The most obvious effect of the narrow doorway configuration was that the transition to ventilation limited burning occurred at a lower HRR than the full doorway configuration. This transition is evident by the reduced oxygen and increased carbon monoxide volume fractions as well as the appearance of flames outside the doorway. The magnitude and front to rear variation of the gas temperatures and species volume fractions were similar for the narrow (test #6) and full door tests using natural gas. Test #5 with heptane showed an interesting reversal of the internal structure of the fires. In contrast to the full door tests, the narrow doorway heptane test showed higher temperatures and CO volume fractions at the rear sample location. Because the narrow doorway testing was limited to only two fuels, further work is needed to generalize the effects of doorway geometry on the compartment fire structure.

4.3 Construction Material Comparisons

In Section 2.1.1, the two different construction materials, Marinite I and Kaowool M-board, were discussed along with the variation in the techniques used to line the enclosure with each material. There were concerns that the different composition of the materials as well as the possibility of leaks through additional seams (M-board) would cause differences in fire behavior. It was conceivable that the higher organic fraction of the Marinite could lead to additional heat release rate and gaseous products of combustion and pyrolysis. It was already known that Marinite experiences significant shrinkage when exposed to 1000 °C temperatures for tens of minutes due to baking off of organics and water.

In order to allay any concerns and reveal any noteworthy effects, experiments with the same fuels and HRRs were repeated for the enclosure lined with each of the two construction materials. For natural gas, the experiments with Marinite were test #1, test #2, and test #3 while test #6.5 used M-board. For heptane (using the same burner), the experiment with Marinite was test #4 and that with M-board was test #7.

One potential impact of construction material was the evolution of combustible organic components. Kaowool M-board has from 3 % to 6 % organic components. Taking into account the 0.23 m^3 volume of the board used to construct the innermost lining, the reported density of 272 kg/m^3, a generic heat of combustion of 35 MJ/kg, and estimating the organic component at 4.5 %, there was about 100 MJ of energy available in the M-board layer. In order for this to impact the calorimetry, the organic material would have to vaporize and enter into the interior of the enclosure where it could be oxidized by enclosure air or by external entrained air as the doorway plume rose into the hood.

To see if the organic components of the M-board added to the HRR, the ratios of the calorimetrically measured HRRs to those estimated from metering of the fuel were calculated for test #6.5 which was the first test conducted with M-board. This ratio was 1.27 for the initial fire size of 76 kW. For the second fire size of 399 kW, the ratio dropped to 1.06. It's worth noting that 27 % of 76 kW and 6 % of 399 kW both result in 20 kW to 25 kW higher calorimeter HRRs than the HRRs calculated from the fuel. The third and fourth fire sizes, 269 kW and 179 kW, respectively, resulted in ratios of about 1.01. The last fire size of 74 kW had a ratio of 1.15 representing a calorimeter HRR 11 kW higher than the fuel metered HRR. Since the uncertainty on the 3 m calorimeter HRR is on the order of 15 %, these differences between calorimeter and fuel metered HRRs are significant, but not easily quantifiable. The conclusion from a review of this data is that there was likely additional heat release from evolved organic components of the construction material that may have been on the order of 20 kW to 25 kW. The effect was more significant for early HRRs during the first "bake out" test and diminished by about half by the end of the test.

Another potential impact of the construction materials is on enclosure temperatures. Gas temperatures could be affected by additional HRR from bake-out of organics and by the thermal conductivity and emissivity of the wall material. Surface temperatures could also be affected by the insulating properties, heat capacities, and emissivities of the materials. Table 21 lists each material's properties including density, thermal conductivity, specific heat, and emissivity. The M-board had significantly lower thermal conductivity and density than Marinite I. Aspirated thermocouples were not deployed for long enough periods in test #6.5 to be useful so only the gas temperatures from heptane fire tests #4 and #7 may be compared. Nominal upper layer gas temperatures in the rear were consistently about 25 °C higher for the M-board than for the Marinite I. Nominal upper layer gas temperatures in the front varied from zero to about 80 °C higher for the M-board. No measurable effect was observed on the lower layer gas temperatures. The observed correlation between upper layer gas temperature and wall material was relatively weak given the relatively large uncertainties in the temperature measurements.

Examination of the total heat flux to the floor for tests with natural gas and heptane showed the nominal heat fluxes for M-board construction were substantially higher (20 kW/m^2 to 30 kW/m^2 for HRRs greater than 200 kW) than for Marinite I construction except for the front heat fluxes for the natural gas tests which were about the same. The larger heat fluxes and temperatures observed for the test using M-board are likely due to better insulating properties compared to Marinite I.

Finally, since mixing and reaction rates and products are influenced by the thermal environment, material differences have some potential to affect gas species measurements. However, an inspection of O_2, CO, CO_2 did not reveal any consistent differences related to construction material.

While the impact of construction materials on the enclosure fire dynamics and measurements was of concern, the primary reasons for trying alternate materials were durability and ease of construction. Table 21 lists some differences between the two construction materials with regards to these and other issues. Of the many differences, the most important is that the M-board survived longer than Marinite I with no cracking and a tolerable amount of warping when it was supported by a sufficient number of furnace pins, especially near the seams. The main differences in the materials are in shrinkage and water content.

Figure 64 through Fig. 69 photographically show some of the differences listed in the table. Figure 64 shows the Marinite I construction before test #2 (natural gas) after experiencing just one fire with a max HRR of 180 kW. The fit is still good, there is no apparent warping, and cracking is minimal with some appearing in the upper right corner. Figure 65 shows a similar view after test #2, but significant cracking has occurred along with some sagging and warping. Figure 66 shows another view after test #6 (natural gas) with even more severe cracking and sagging and a gap from a fallen piece of inner lining.

Figure 67 through Fig. 69 show photographs of M-board construction. Figure 67 shows a close-up view of the furnace pins and tight seams before the enclosure was exposed to a fire. Figure 68 shows a similar view after the first fire, test #6.5. There is a small amount of warping at the wall seam. Figure 69 shows the whole enclosure interior with the back removed. Extensive warping and sagging occurred, but not much cracking. The warping and sagging were remedied for the construction of the next enclosure lining by increasing the number of furnace pins, especially near the seams. Figure 70 depicts an aspirated thermocouple probe and furnace pins that have been highly degraded by extremely high heat fluxes and temperatures. The pins still held, but were extremely fragile and easily broken in this state.

Table 21. Comparison of thermal, physical and construction properties of wall lining materials.

Propery	Marinite I	Kaowool M-board
Sheet size	1.2 m × 2.4 m (4 ft × 8 ft)	0.9 m × 1.2 m (3 ft × 4 ft)
Number of internal seams	12	18
Fastener type	Screws, washers	Furnace pins
Fastener durability	Survive well	Survive with some warping
Cracking	Some after first test, noticeable increases with each test	No cracking
Warping	Increases with each test up to sagging of several cm	Was significant near seams, not significant with additional fasteners
Fragility	Still stiff and strong after multiple tests although less securely attached and susceptible to falling loose	Fairly soft and fragile even when new, should not be handled after exposed to fires
Shrinkage @ 982 °C	13.2 %	2.2 %
Thermal Conductivity, k	0.11 W/m·K to 0.13 W/m·K (24 °C to 538 °C)	0.06 W/m·K to 0.22 W/m·K (0.20 W/ m·K at 1000 °C)
Density, ρ	737 kg/m^3	272 kg/m^3
Specific Heat, c_p	1172 J/kg·K to 1424 J/kg·K (93 °C to 425 °C)	Not specified
Emissivity, ε	0.74±0.4	0.95
Organic Content	4 % to 8 %	3 % to 6 %
Water Content	3 % of dry weight	< 0.5 %
Ease of Cutting	Slow, requires saw, produces a lot of dust	Can be cut with a utility knife.

Figure 64. Photo of rear gas sample location and Marinite I construction before test #2.

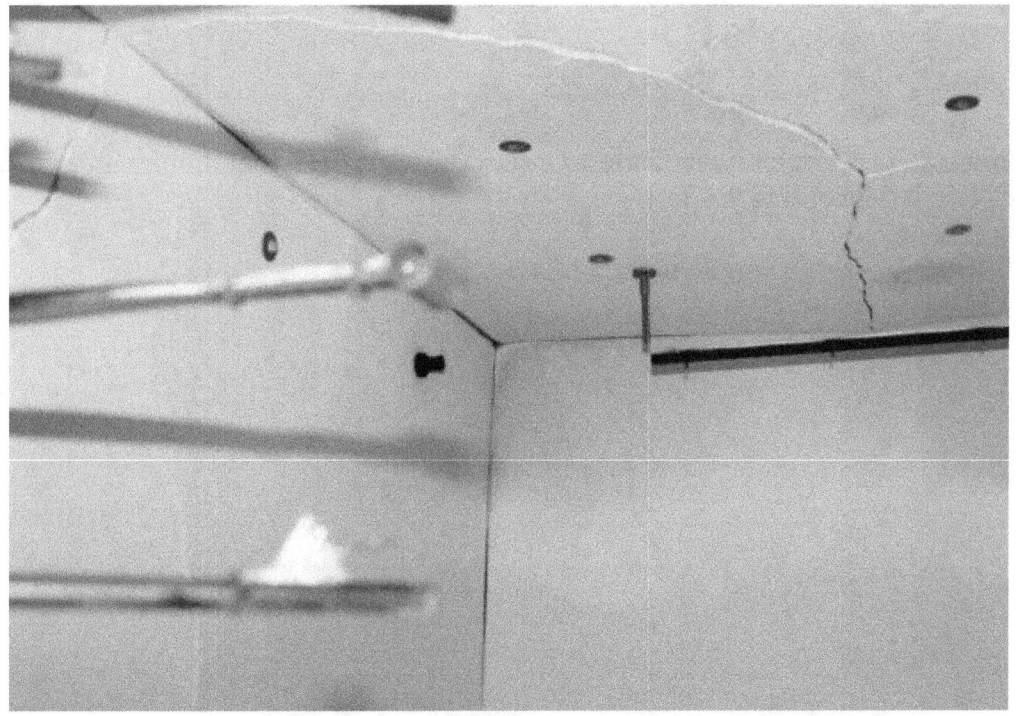

Figure 65. Photo of rear gas sample location and Marinite I construction after test #2 with natural gas.

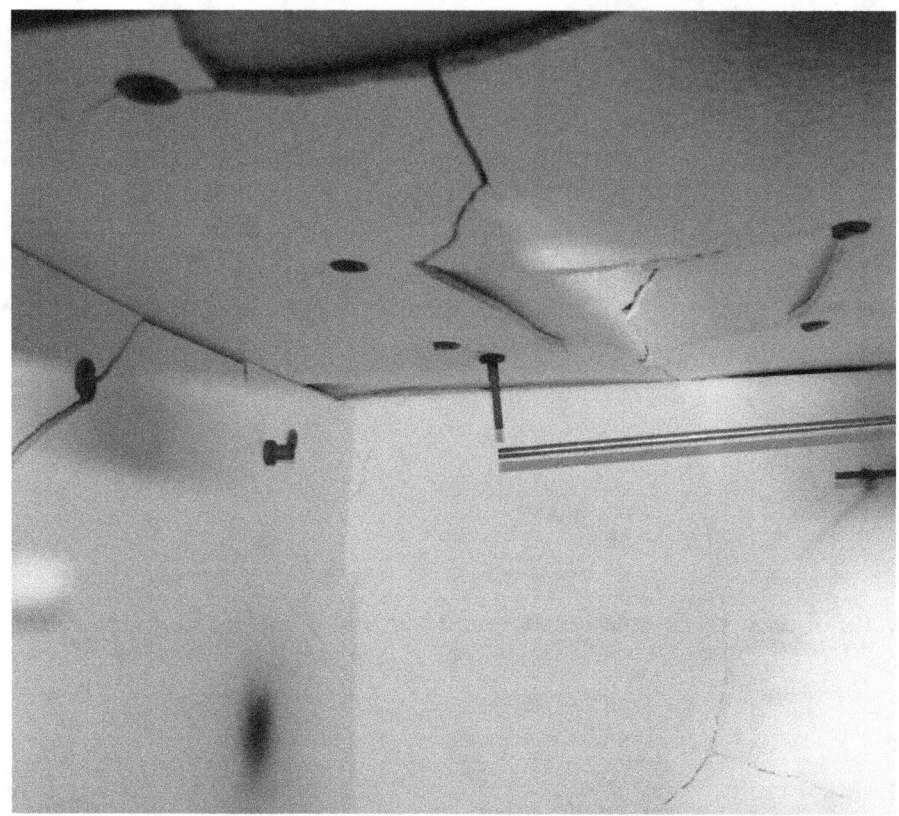

Figure 66. Photo inside enclosure after test #6 showing Marinite I condition. Large chunks of ceiling fell to the floor during this test.

Figure 67. Photo of rear gas sample location showing Kaowool M-board and furnace pin construction prior to test #6.5.

Figure 68. Photo of rear gas sample location after test #6.5 with natural gas using Kaowool M-Board.

Figure 69. Photo of enclosure with rear wall removed after test #10 using toluene. Notice melted stainless steel furnace pins and bowed M-board wall but very little cracking of walls.

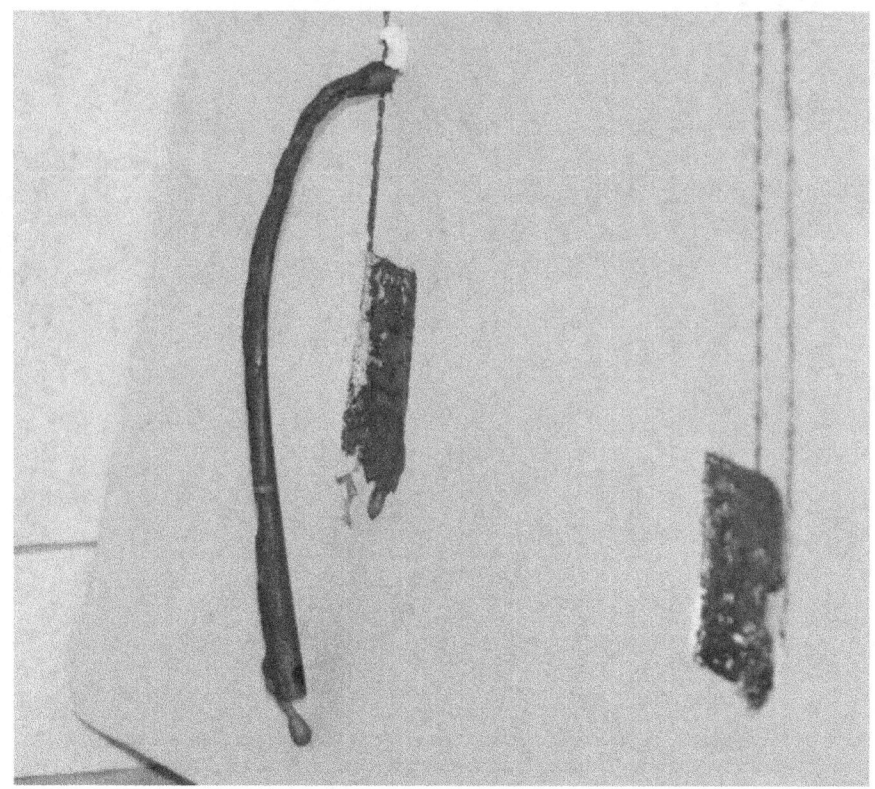

Figure 70. Photo of rear lower aspirated probe (TR24A) after test #16 with polystyrene. Stainless steel furnace pins melted and dripped along the wall.

4.4 Burner Type Comparisons

A number of different burner designs were used during this test series for both practical and technical reasons. Natural gas was delivered using square gravel-filled burners (A and B), shown in Fig. 2. Although the surface areas of these two burners varied by almost a factor of 4, and only burner B was water cooled, there was no measurable difference in the fire conditions inside the enclosure as a function of HRR.

During the original test series in the 1990s, heptane was burned in a round pan and the depth of the pool was controlled. It was found that steady underventilated burning could not easily be attained using this method. In the test described here, liquid fuels were delivered using a square water-cooled pool burner with inclined walls (burner B) and a spray nozzle directed into a round catch pan (burner C). Burner B performed well in attaining steady burning for the heptane and toluene fuels (test #7 and test #10). In these tests, a large amount of radiation from the upper layer was imposed on the fuel surface. This additional heat flux was necessary to reach underventilated conditions since open burning tests with heptane using burner B resulted in a maximum HRR of only about 80 kW (the fuel surface area was 625 cm^2). During test #8 and test #9 attempts were made to achieve underventilated fires using methanol and ethanol liquid fuels with burner B. These attempts failed because of lack of re-radiation from the upper layer to increase the burning rate necessary for ventilation limited conditions. This was due to the combined effect of low heat of combustion and lack of soot to act as a radiation source. The maximum HRR was 20 kW and the maximum heat flux to the floor was 3 kW/m^2 for the alcohol fuels using burner B. The spray burner (C) allowed a greater burning area, enabling the attainment of underventilated fire conditions for the alcohol fuels in test #11 and test #12.

Test #15 was conducted with heptane fuel to compare the spray burner performance with the results of test #4 and test #7 using the heptane pool burner B. Significant differences were observed in the species volume fractions, temperatures and heat fluxes at the same HRR using different burners. The spray burner C appeared to be less efficient both locally at the interior sampling locations and globally in the exhaust stack. During the 375 kW steady heptane fire in test #15, the rear sample location CO, THC and soot measurements were between 3 and 5 times greater than the same fire size using the pool burner in test #4. The same trend was observed in the exhaust stack species measurements. The interior gas temperatures and heat flux to the floor were lower for the spray burner test. One possible explanation for the inefficiency of the spray burner is that the fuel was injected higher in the compartment where the differences in temperature and oxygen volume fraction could lead to differences in mixing, impacting the compartment fire dynamics. This effect should be considered when making comparisons of different fuel types in the results presented here. Specifically, the CO volume fractions observed in test #11 and test #12 using the alcohols fuels could have been enhanced by the configuration of the fuel delivery (burner C).

5 Analysis of Compartment Chemistry

The dynamics and chemistry of the processes occurring in a compartment fire are very complex. In an attempt to better understand the chemistry of the compartment fire experiments, gas species and soot measurements were made at two locations in the hot upper layer of the compartment and far downstream in the exhaust stack, where the temperatures had cooled, typically to less than 200 °C.

5.1 Mixture Fraction

It is useful to consider the compartment fire composition measurements in terms of the mixture fraction. The use of mixture fraction to analyze flame data was first used by Bilger [39] and later modified by Peters [40] and others. The mixture fraction approach has been widely used to represent the chemistry in turbulent flame models and fire field models, and has been used to analyze the structure of laminar counterflowing and coflowing hydrocarbon and alcohol flames [41,42].

Pool fires and compartment fires differ from simple laminar flames, as they are typically transient and turbulent by nature. Yet, application of the mixture fraction concept to these complex combustion situations can provide additional insight into the structure of the fire. The mixture fraction approach allows evaluation of a set of species measurements in terms of self-consistency, and at the same time facilitates rapid assessment of the overall behavior of a combustion system. Floyd et al [43] applied the mixture fraction approach to evaluate the species composition at various locations in compartment fires. Pitts [17] measured the local equivalence ratio at various locations in compartment fires, investigating the possibility of a correlation for CO. Since there is a one-to-one correspondence between mixture fraction and equivalence ratio, the approach used here is similar to that used previously by Pitts [17] and other experimentalists, with the difference that soot is considered in the analysis of mixture fraction and local equivalence ratio.

Sivathanu and Faeth [44] considered the relationship between soot and mixture fraction in an effort to improve the understanding associated with radiative emissions from fires. Their measurements [44] clearly showed that soot did not correlate well with mixture fraction in laminar hydrocarbon diffusion flames. Their data suggest, however, a relationship between soot volume fraction and temperature in the fuel rich regions of turbulent hydrocarbon diffusion flames.

In this study, the mixture fraction was used to evaluate the species composition at various locations in the hot upper layer of the compartment for a number of reasons. First, the analysis provides a check on the quality of the data and provides insight into the chemistry of compartment fires. Second, the significance of the inclusion of soot as part of the mixture fraction analysis was investigated. The intent of this part of the study is to determine if the inclusion of soot adds coherence to the mixture fraction approach. Finally, the importance of measurement uncertainty is highlighted, and its value is quantified as part of the mixture fraction analysis.

Definition of Mixture Fraction

The mixture fraction is a non-dimensional quantity representing the mass fraction of a species, at a particular location, that was originally part of the fuel stream. The mixture fraction based on carbon containing species is defined as follows:

$$Z = Y_F + Y_{co} \frac{MW_F}{x\,MW_{co}} + Y_{co_2} \frac{MW_F}{x\,MW_{co_2}} + Y_{Soot} \frac{MW_F}{x\,MW_{Soot}} \tag{15}$$

where MW_i is the molecular mass of chemical species i, Y_i is the mass fraction of that species, x is the number of carbon atoms in the parent fuel molecule ($C_xH_yO_z$), MW_F is the molecular mass of the parent fuel, MW_{CO} is 28 g/mol, MW_{Soot} is taken as 12 g/mol (assuming that soot can be approximated as pure carbon), and MW_{CO2} is 44 g/mol. Alternative definitions of mixture fraction yield results similar to those shown below.

In the experiments reported here, the measurement of total unburned hydrocarbons was made using the total hydrocarbon analyzer, reported on an equivalent methane basis. The subscript F in the first term of Eq. 15 can be thought of as referring to total hydrocarbons (THC).

In the fire literature, soot is typically not considered in Eq. 15. Here, it is included formally. But in the analysis given below, this term is initially neglected, because it is small. Its inclusion is important for highly sooting conditions, as will be shown in the results section below.

The mass fraction, Y_i, of each species i is determined from the measured volume fraction, X_i, by the following expression:

$$Y_i = X_i MW_i / MW_{tot} \tag{16}$$

MW_{tot} represents the average molecular mass of all gas species and is a function of the local composition.

$$MW_{tot} = \sum_i X_i MW_i \tag{17}$$

The state relations can be derived by considering the idealized reaction of a hydrocarbon fuel, rewritten here in an expanded form of Eq. 9:

$$C_xH_yO_z + \eta(x+y/4-z/2)(O_2 + 3.76N_2) \rightarrow \max(0,1-\eta)C_xH_yO_z + \min(1,\eta)xCO_2$$
$$+ \min(1,\eta)(y/2)H_2O + \max(0,\eta-1)(x+y/4-z/2)O_2 + \eta(x+y/4)3.76N_2 \tag{18}$$

where the function $\max(\alpha,\beta)$ returns the larger of the two parameters, α or β, and the function $\min(\alpha,\beta)$ returns the smaller of the two parameters, α or β. Here, η is a parameter ranging from zero (all fuel and zero oxygen) to infinity (all oxygen and zero fuel) and becomes unity for stoichiometric conditions. The definition of η shows that it is the reciprocal of the local fuel equivalence ratio, ϕ.

$$\phi = \frac{(F/A)}{(F/A)_{st}} = \frac{MW_F / \eta(x+y/4-z/2)(MW_{O_2} + 3.76MW_{N_2})}{MW_F /(x+y/4-z/2)(MW_{O_2} + 3.76MW_{N_2})} = \frac{1}{\eta} \tag{19}$$

where F/A is the fuel-air ratio and the subscript st refers to stoichiometric conditions. The idealized mass fractions of products are obtained from the right side of the Eq. 18. For stoichiometric conditions, $Y_F = Y_{CO} = 0$, and Eq. 15 leads to:

$$Z_{st} = Y_{CO2} \frac{MW_F}{x\, MW_{CO2}} \tag{20}$$

The value of the stoichiometric mixture fraction for the fuels considered in this report is shown in Table 22. Its value varied from about 0.0554 for natural gas to 0.1346 for methanol.

Table 22. Stoichiometric value of the mixture fraction (Z_{st}) for different fuels.

Fuel	Chemical Formula	Z_{st}
Methane	CH_4	0.0552
Natural Gas	0.93 CH_4 + 0.04 C_2H_6 + 0.01 C_3H_8 + 0.01 CO_2 +... [*]	0.0554 ± 0.0002 [**]
n-Heptane	C_7H_{16}	0.0622
Toluene	C_7H_8	0.0694
Polystyrene	$(C_8H_8)_n$	0.0705
Methanol	CH_3OH	0.1346
Ethanol	C_2H_5OH	0.1006
* typical composition; actual composition varies day to day.		
** average value based on measured natural gas composition.		

A mixture fraction calculation for a methane-air flame is presented here as an example. For methane, Eq. 18 becomes:

$$CH_4 + 2\eta(O_2 + 3.76N_2) \rightarrow \max(0, 1-\eta)CH_4 + \min(1, \eta)CO_2 + 2\min(1, \eta)H_2O$$
$$+ 2\max(0, \eta-1)O_2 + 7.52\eta N_2 \tag{21}$$

The traditional mixture fraction model holds that the mass fraction, Y_i, of products can be determined through the right side of Eq. 21 as follows:

$$Y_{CH_4} = \max(0, 1-\eta)MW_{CH_4}/MW_{tot}$$
$$Y_{CO_2} = \min(1, \eta)MW_{CO_2}/MW_{tot}$$
$$Y_{H_2O} = 2\min(1, \eta)MW_{H_2O}/MW_{tot} \tag{22}$$
$$Y_{O_2} = 2\max(0, \eta-1)MW_{O_2}/MW_{tot}$$
$$Y_{N_2} = 7.52\eta MW_{N_2}/MW_{tot}$$

where Y_{CO} and Y_{Soot} in Eq. 18 were taken as zero for this mixture fraction model calculation. The molecular mass of the mixture is a function of the local composition and can be calculated from the reactant concentrations:

$$MW_{tot} = MW_{CH_4} + 2\eta(MW_{O_2} + 3.76MW_{N_2}). \tag{23}$$

Since Y_{CO} and Y_{Soot} are assumed to be equal to zero and $Y_F = Y_{CH_4}$ the mixture fraction defined in Eq. 15 can be rewritten as:

$$Z = Y_{CH_4} + Y_{CO_2} \frac{MW_{CH_4}}{x\, MW_{CO_2}} \tag{24}$$

Using Eqs. 22 and 23, Eq. 24 can be rewritten as:

$$Z = \frac{MW_{CH4}}{MW_{tot}} = \frac{MW_{CH4}}{MW_{CH4} + 2\eta(MW_{O_2} + 3.76MW_{N_2})} \, , \qquad (25)$$

and

$$\eta = \frac{(1-Z)}{Z} \frac{MW_{CH4}}{2(MW_{O_2} + 3.76MW_{N_2})} = \frac{(1-Z)}{Z}(F/A)_{st} \, . \qquad (26)$$

Figure 71 presents the relationship between the mixture fraction and the equivalence ratio ($1/\eta$) as delineated in Eq. 26 for the methane-air system. Under stoichiometric conditions ($\eta = 1$), the mixture fraction is 0.0552 for a methane-air flame as listed in Table 22. In Fig. 71, natural gas is treated as if it were methane. The figure shows that the mixture fraction compresses a large range of equivalence ratio values. Figure 72 shows the relationship between the mass fraction and the mixture fraction for most of the major species in the methane-air system, when Y_{CO} is taken as zero.

Mixture Fraction Uncertainty

The uncertainty in the mixture fraction is propagated through Eq. 15 and is based on the measurement uncertainty of the species concentrations. The positive square root of the estimated variance, $U_Z(Y_i)$, is obtained from

$$U_Z^2(Y_i) = \sum_{i=1}^{N} \left[\frac{\partial Z}{\partial Y_i} \right]^2 U_{Y_i}^2 \qquad (27)$$

where U_{Y_i} is an estimate of the combined expanded measurement uncertainty of the measured mass fraction, Y_i, of species i.

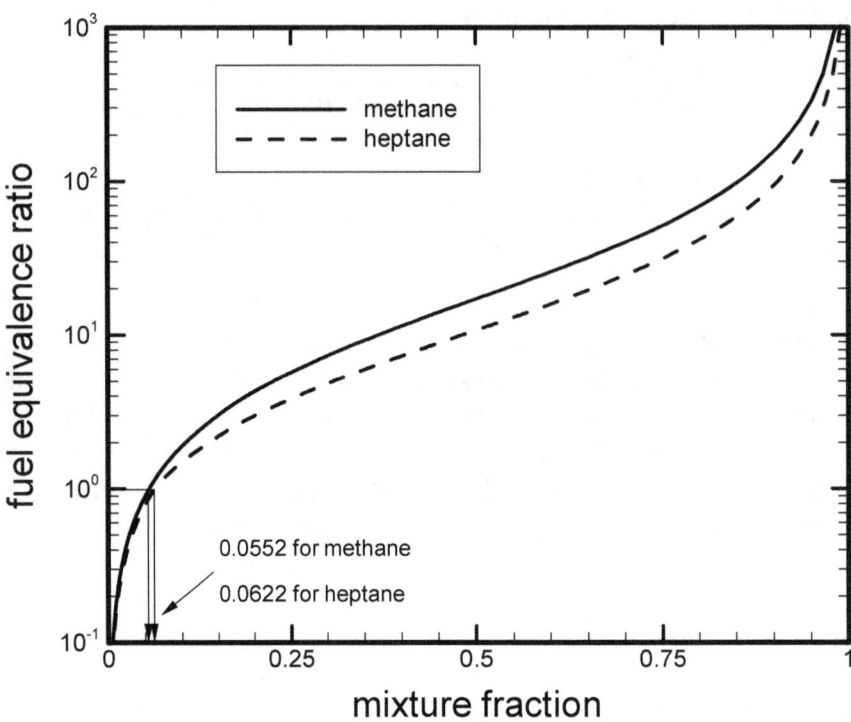

Figure 71 The equivalence ratio as a function of mixture fraction for nonpremixed flames burning methane and n-heptane.

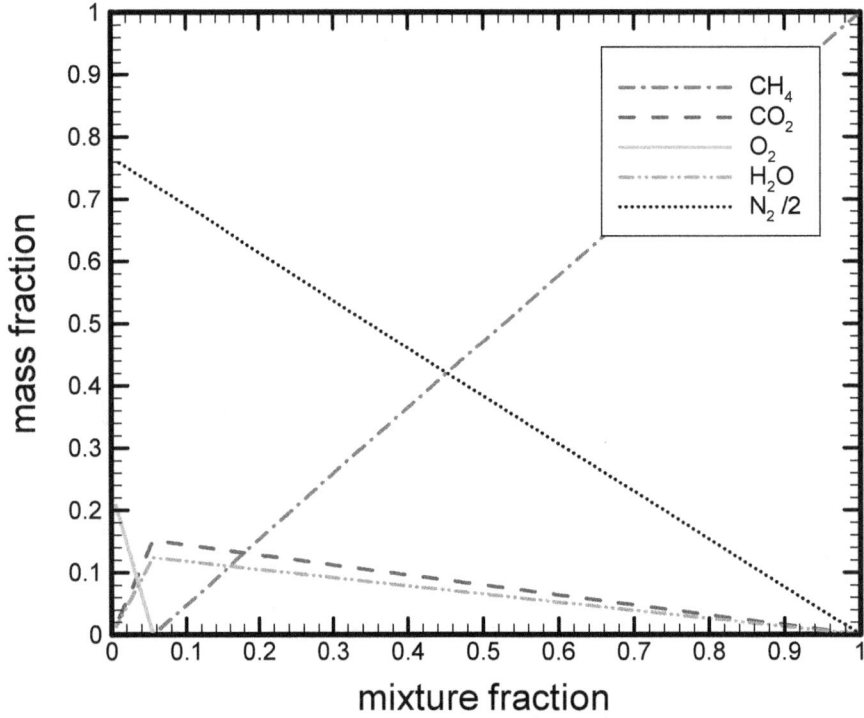

Figure 72 The mass fraction vs. the mixture fraction calculated by the single-parameter mixture fraction model.

Species Composition Results in terms of Mixture Fraction

In this section, the time-varying species measurements are presented as a function of the mixture fraction. The results are organized in terms of fuel type, since the fuel type establishes the basis for the correlation (see Eqs. 9 and 15).

The species data are considered in terms of the species mass fraction (Y_i), which is plotted as a function of the local mixture fraction (Z), based on the fuel mass. Measurements from the front and rear of the compartment, for all fire conditions (i.e., heat release rate, burner type) and all times during the experiment are plotted on a single graph in terms of mixture fraction. The mass fractions of H_2O and N_2 were not measured in the experiments; the values of these species in this report (and shown in Fig. 73 through Fig. 96) are estimated from the stoichiometric relation (Eqs. 9-11). The mass fractions of the unburned hydrocarbons (THC) in each plot were taken from the hydrocarbon analyzer measurements. The total hydrocarbons (THC) results were normalized in terms of the equivalent fuel molecule for each fuel type.

The lines in Fig. 73 through Fig. 77, Fig. 80, and Fig. 81) represent, respectively, complete stoichiometric combustion and the hypothetical case when only CO_2 is produced (no CO or soot; see Fig. 72). In some cases, because of the number of data points, the theoretical lines are somewhat obscured. The lines on the average steady-state measurement results are easier to distinguish as the plots are less crowded. In those plots, the propagated uncertainty is also presented. Soot was not measured at all times, but only during the periods when the fire heat release rate was quasi- steady. Thus, soot is shown only on the plots labeled "(b)" and is presented with the time-averaged gas species results only. The data labeled "THC" in the figures represents the total unburned hydrocarbons measured with the FID detector on the total hydrocarbon analyzer.

Natural Gas

Figure 73 presents all of the gas species measurements taken during all of the natural gas experiments (tests #1 - #3, and #6) in both the front and rear of the compartment as a function of mixture fraction. Figure 73a shows all of the transient measurements for all of the natural gas tests with the full-door configuration (tests #1 - #3 and #6). Figure 73b shows the time-averaged steady-state measurements (and represents the same data as shown in Fig. 41). At any single location, the mixture fraction can vary from lean to rich, due to the dynamics of the fire. The stoichiometric mixture fraction (Z_{st}) is a useful reference point for consideration of fire chemistry (see Table 22; $Z_{st} = 0.0544$). For fuel lean conditions ($Z < Z_{st}$), the measured mass fractions of methane and carbon monoxide are near zero. As the mixture fraction increases, the mass fraction of oxygen decreases, and the carbon dioxide and water vapor mass fractions increase. For mixture fraction values greater than stoichiometric, the oxygen mass fraction approaches zero, whereas the fraction of unburned fuel increases approximately linearly. Under these conditions, the generation of carbon monoxide is observed and Y_{CO} attains a maximum value of about 0.04 g/g.

As seen in the figure, the hypothetical lines show reasonable agreement with the measurements for fuel lean and near-stoichiometric conditions. As the mixture fraction increases beyond stoichiometric, however, the difference between the hypothetical lines and the measurements becomes considerable. The value of Y_{CO} is not negligible for fuel rich conditions. As a result, the hypothetical lines over-predict the CO_2 mass fraction by about 10 % for mixture fraction values for $Z_{st} < Z < 0.2$. As expected, the plots show that the simple traditional mixture faction approach does not correlate the experimental results for CO. This behavior is also observed in laminar flames, which is attributed to finite rate chemistry effects associated with slow CO chemistry [41]. Other approaches to predict CO, possibly using variables that are functions of mixture fraction, will need to be considered to improve predictions of its concentration.

The vertical and horizontal error bars in Fig. 73b represent the combined expanded uncertainty of the mass fractions of gas species and the mixture fraction, respectively. For $Z < 0.2$, the uncertainties are relatively small, and the mixture fraction correlations show reasonable agreement with the experiments. For $Z > 0.2$ (in Fig. 73b), the uncertainties in the mixture fraction and the mass fraction of gas species, especially unburned fuel and nitrogen, become large, and the maximum relative error of mixture fraction reaches values as large as 15 %. Typically, the mixture fraction model is within experimental uncertainty for the conditions considered in this study, when natural gas was the fuel. The results for H_2O are the exception to this. Direct measurement of the H_2O concentration would be helpful in this regard. Since the H_2 concentration can be on the order of a few percent in a hydrocarbon diffusion flame [42], its measurement would also be of interest..

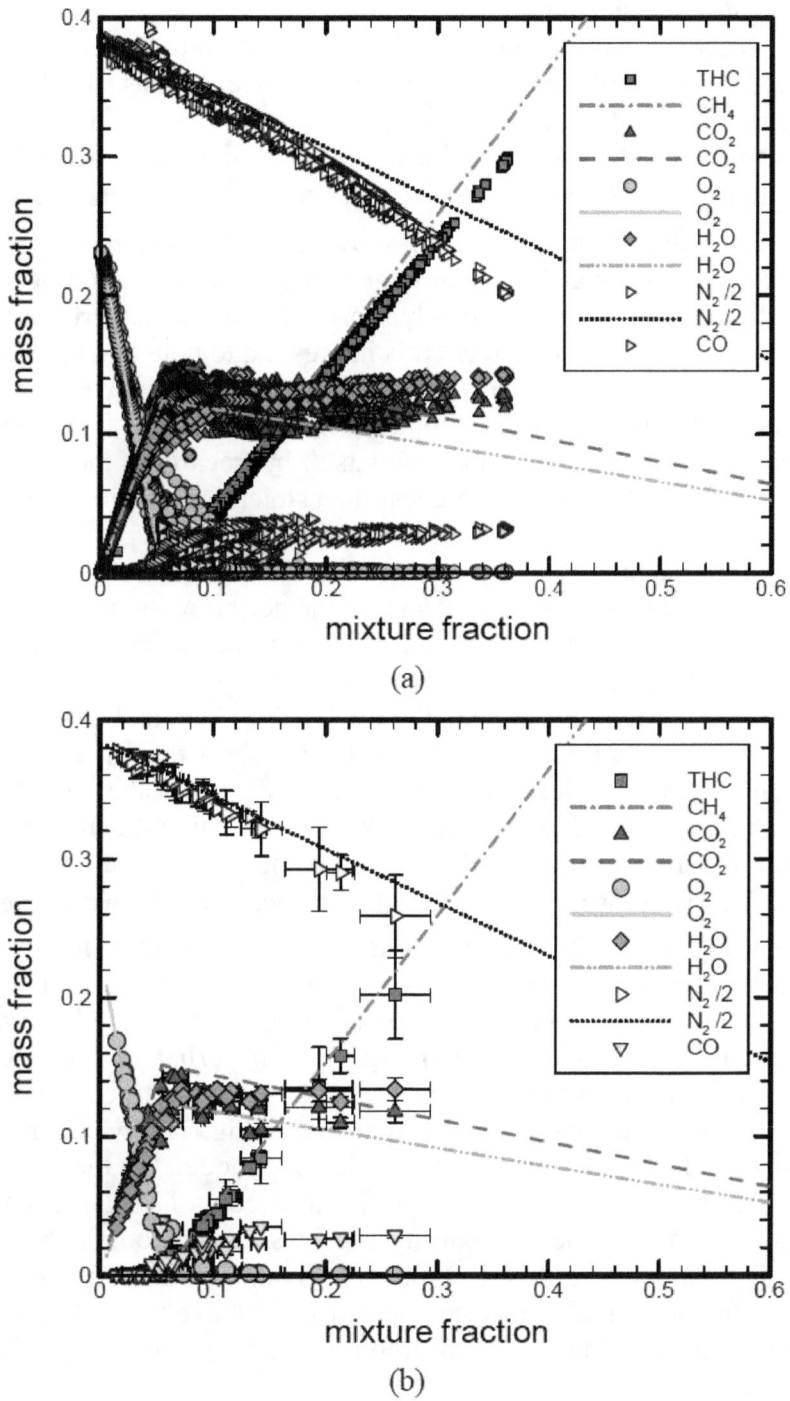

(a)

(b)

Figure 73 Mass fractions of front and rear compartment gas species as a function of mixture fraction for the natural gas fire tests #1-#3, and #6: (a) transient measurements and (b) time-averaged measurements during the period when the HRR was quasi-steady.

Condensed-Phase Hydrocarbon Fuels

Figure 74 to Fig. 76 show the mass fraction as a function of mixture fraction for the fires burning heptane (tests #4, #5, #7, #15), toluene (test #10) and polystyrene (tests #13, #14, #16), respectively. For small values of mixture fraction ($Z << Z_{st}$), the species mass fractions agree with the mixture fraction model for all three fuels. Figure 74 shows, however, for near-stoichiometric conditions, Y_{CO} is non-zero, which leads to Y_{CO_2} lower than predicted by the mixture fraction model. Figure 74b and Fig. 75b show that the measurement uncertainty was relatively small for lean mixture fractions. For large values of the mixture fraction, the variance of the species mass fraction results was relatively broad. This is particularly true for the transient results, but also for the time-averaged results. It is interesting to note that the figures show that the local conditions are not fuel-rich for any of the conditions investigated during the toluene and the polystyrene tests. Although there was a high concentration of mass fraction results about near-stoichiometric conditions, the negligible amounts of hydrocarbons measured during these tests led to mixture fraction values which were less than stoichiometric in value.

Species concentration results in the fire literature, such as those presented in Fig. 74 to Fig. 76, are typically reported without consideration of soot in the definition of mixture fraction. It is correct to include soot in Eq. 15 as the conserved scalar approach is based on the idea that elemental mass is neither created nor destroyed in a fire. The appearance of the plots qualitatively change when soot is considered. Figure 77a, Fig. 77b, and Fig. 77c replot the data shown in Fig. 74b, Fig. 75b, and Fig. 76b, respectively, and show that the inclusion of soot reduces the scatter in the mass fractions for large values of Z, while otherwise leaving the plots unchanged. Inclusion of soot stretches the value of Z proportional to the measured soot mass fraction in a non-linear manner as illustrated in Fig. 78 using the heptane, toluene and polystyrene results. This is because Y_{Soot} is negligible for lean conditions, whereas it is significant for large values of Z, taking on values as large as 0.1. Neglecting soot for the fires burning natural gas and alcohol fuels is reasonable, whereas considering it for heavily sooting fires is necessary. The scatter in the mass fractions was reduced for these fuels when soot was considered in the definition of Z (see Eq. 15). In Fig. 77a, Fig. 77b, and Fig. 77c, the sum of the soot and total hydrocarbons (THC) appears to closely follow the mixture fraction model results. The results plotted in this way are particularly convincing in Fig. 77a, where the independent results for soot and THC do not follow the state relationship model, but their sum does. Interestingly, Fig. 77 shows that there was no significant amount of THC measured in the upper layer of the compartment in the toluene or polystyrene fires. The carbon in the upper layer of these fires is primarily in the form of CO, CO_2 or soot. Examination of Fig. 52 presents the same data, which reaffirms that the total HC measurements were relatively small in the polystyrene and toluene fires, and that the unburned hydrocarbons did not represent a significant fraction of the carbon in the upper layer.

It is also of interest to examine the results in which soot was not used in the mixture fraction definition. This is the typical manner of representing the data in the combustion and fire literature. The value of neglecting soot in the mixture fraction definition is seen in the comparison of Fig. 74 to Fig. 76 with Fig. 77. For conditions when $Z < Z_{st}$, the results that neglect soot in the mixture fraction definition more closely track the state relationship model than the results that consider soot in the mixture fraction definition. The reasons for this are unclear, but this result may be important for understanding the success of mixture fraction correlations without consideration of soot. The comparison of Fig. 74 to Fig. 76 with Fig. 77 illustrates the importance of including soot in the mixture fraction definition, particularly for conditions when $Z>Z_{st}$. Further investigation of these results may prove useful in the development of predictive capabilities for compartment fire species.

Since CO chemistry is relatively slow compared to many flame processes, it may be reasonable to consider the effect of the local temperature on the state relationship model. Figure 79a and Fig. 79b shows the transient values of Y_{CO} and Y_{CO2} in the front and rear of the compartment (same data as Fig. 74a) as a function of mixture fraction with symbols colored to represent the local temperature in the heptane experiments (tests #4, #5, and #7). The results show that while there are some general trends associated with temperature, the values of Y_{CO} are quite scattered and do not systematically correlate with temperature.

Alcohol Fuels

Figure 80 and Fig. 81 present the mass fractions of products for the methanol (tests #8, #12) and ethanol fires (tests #9, #11), respectively. Figure 80 shows the gas species mass fractions at the front and rear of the compartment as a function of mixture fraction for the methanol fire tests #8 and #12. Figure 80a and Fig. 81a shows all of the transient measurements for the methanol and ethanol tests, respectively and Fig. 80b and Fig. 81b shows the averaged quasi-steady measurements for the methanol and ethanol tests, respectively. The soot measurements are plotted with the quasi-steady results, but were not measured and are not shown as part of the transient results. Compared to the mixture fraction model, the concentrations of water vapor are too large, while the unburned fuel and carbon dioxide fall below the predictions. For fires burning both of the alcohol fuels, the value of Y_{CO} is as high as 10 %, which is over two times its value in the natural gas and heptane fires.

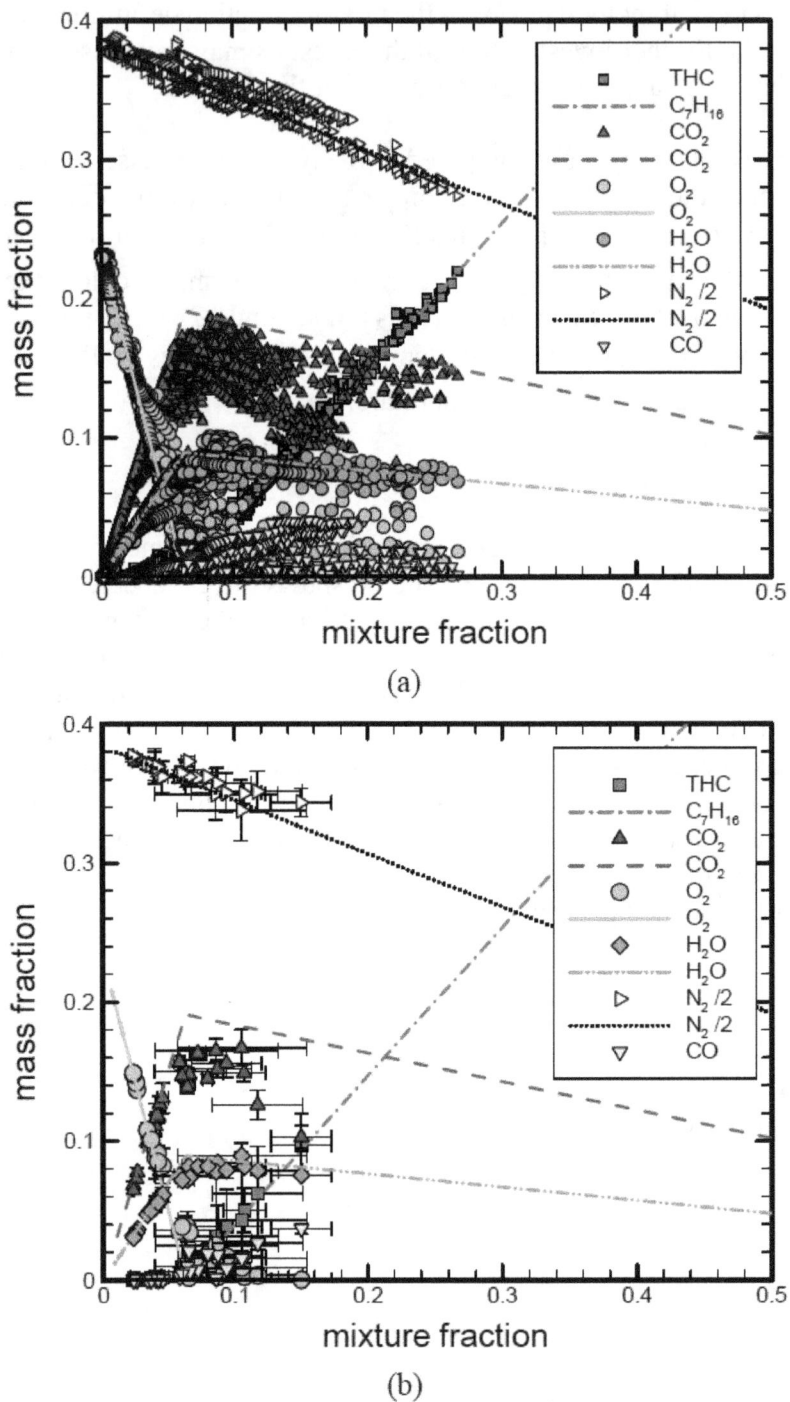

Figure 74 Mass fractions of front and rear compartment gas species as a function of mixture fraction for the heptane fire tests #4, #5, #7 and #15: (a) transient measurements and (b) time-averaged measurements during the period when the HRR was quasi-steady.

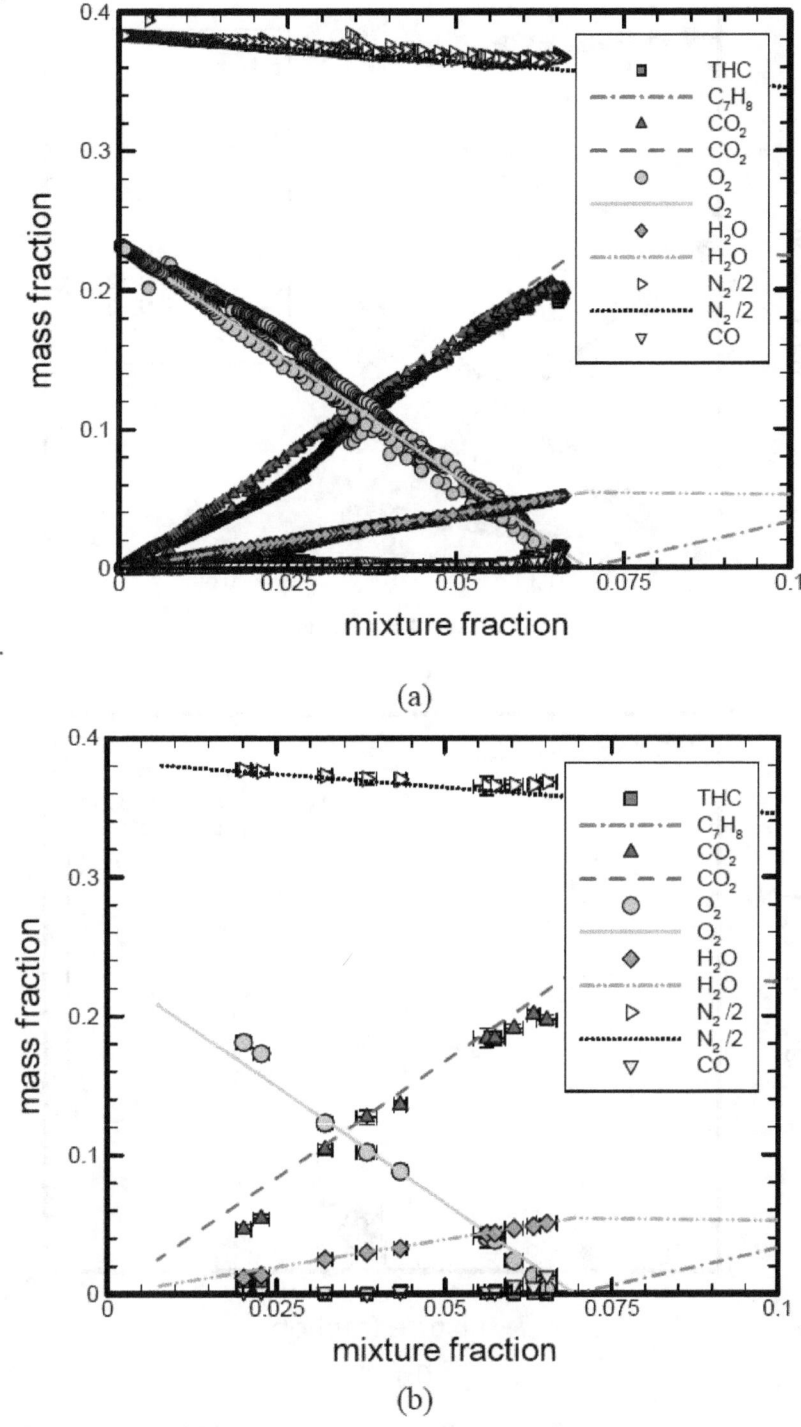

(a)

(b)

Figure 75 Mass fractions of front and rear compartment gas species as a function of mixture fraction for the toluene fire test #10: (a) transient measurements and (b) time-averaged measurements during the period when the HRR was quasi-steady.

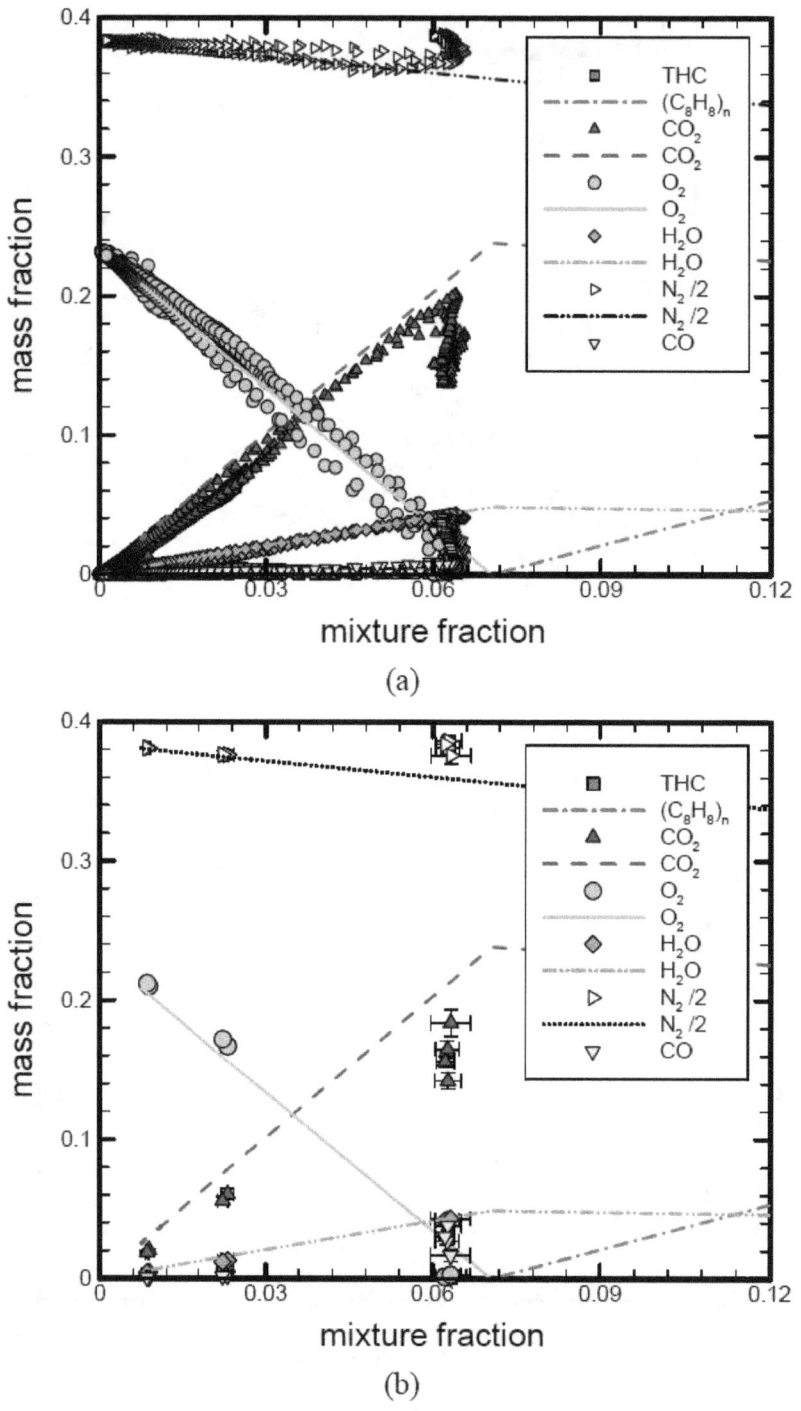

Figure 76 Mass fractions of front and rear compartment gas species as a function of mixture fraction for polystyrene fire tests #13, #14, and #16: (a) transient measurements and (b) time-averaged measurements during the period when the HRR was quasi-steady.

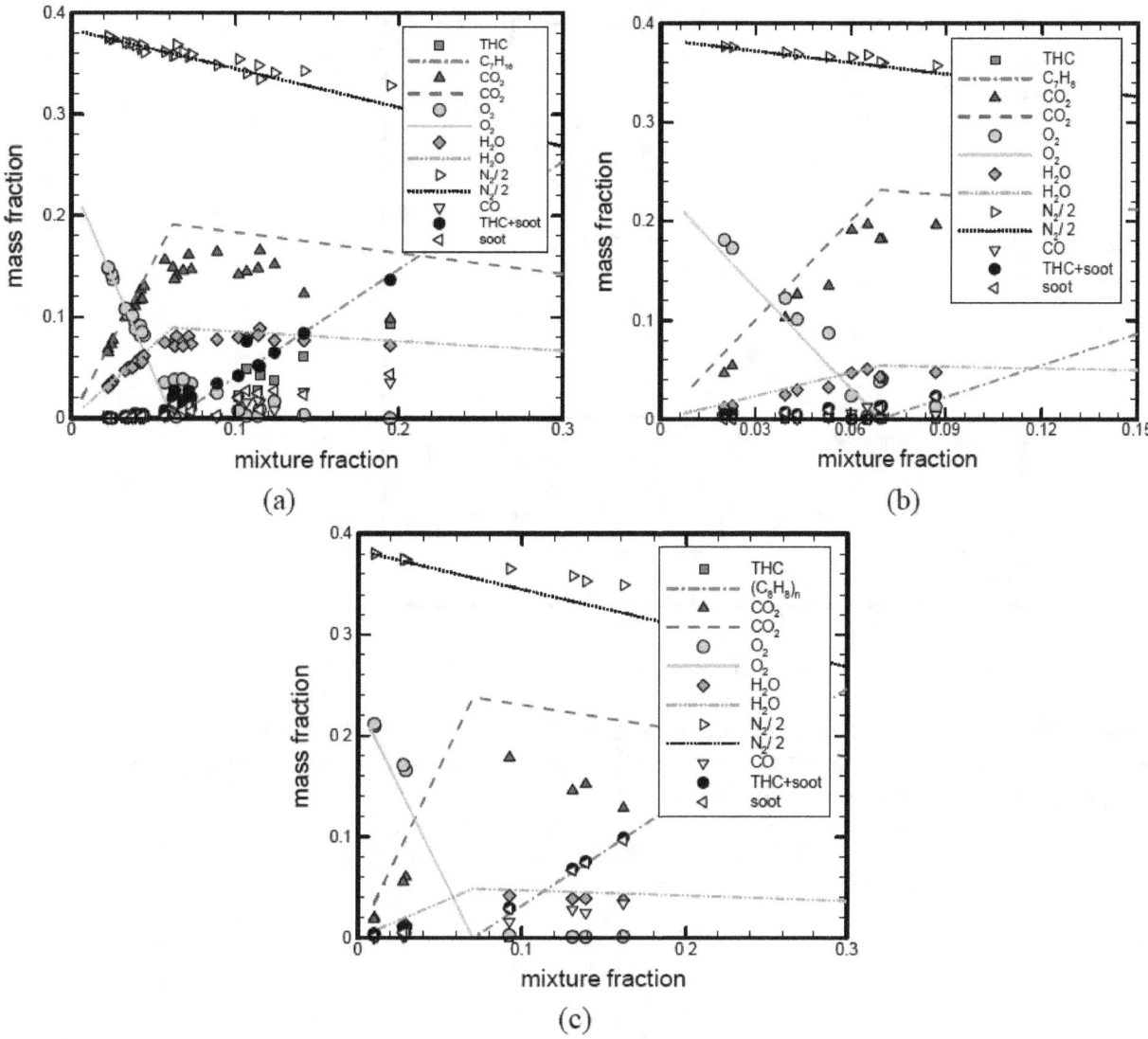

Figure 77 Mass fractions of front and rear compartment gas species as a function of mixture fraction (with soot included) for the time-averaged measurements when the HRR was quasi-steady in the experiments burning (a) heptane, (b) toluene, and (c) polystyrene.

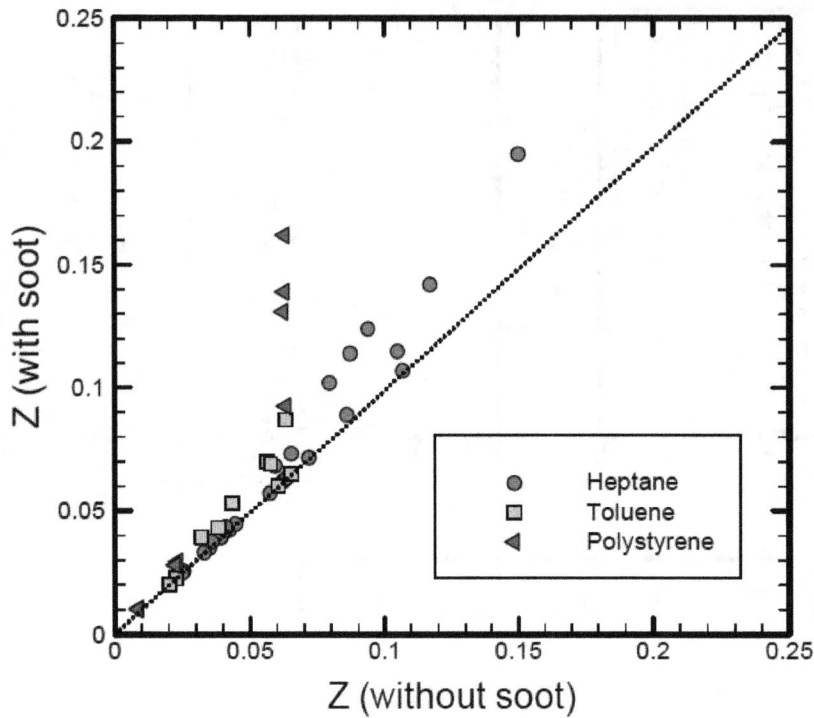

Figure 78 Comparison of mixture fractions calculated with and without soot using the time-averaged species measurements when the HRR was quasi-steady in the experiments burning (a) heptane, (b) toluene, and (c) polystyrene.

.

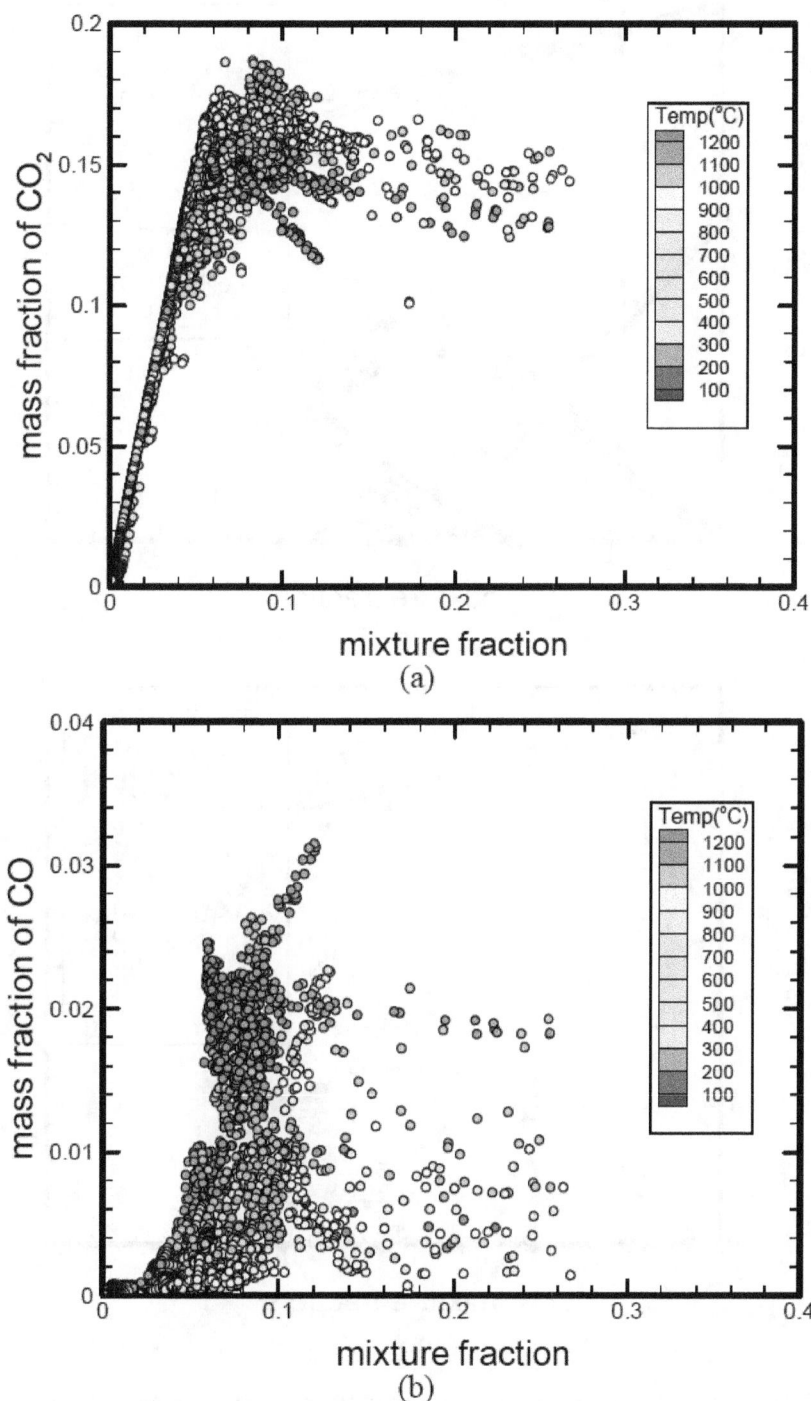

Figure 79 Transient values of (a) Y_{CO_2} and (b) Y_{CO} in the front and rear of the compartment as a function of the mixture fraction (without soot) with symbols colored to represent the local temperature for the heptane fire tests #4, #5, and #7.

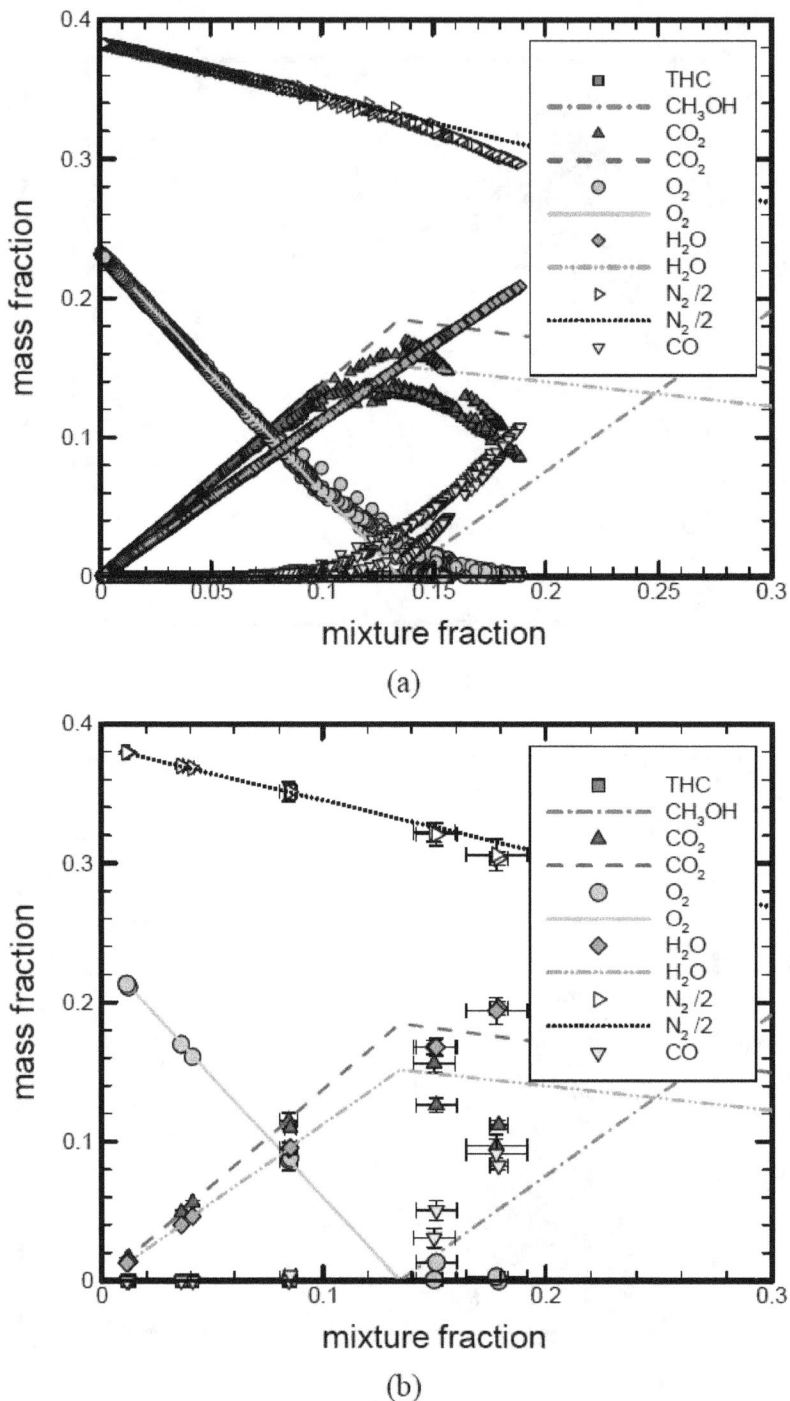

(a)

(b)

Figure 80 Mass fractions of front and rear compartment gas species as a function of mixture fraction for methanol fire tests #8 and #12: (a) transient measurements and (b) time-averaged measurements during the period when the HRR was quasi-steady.

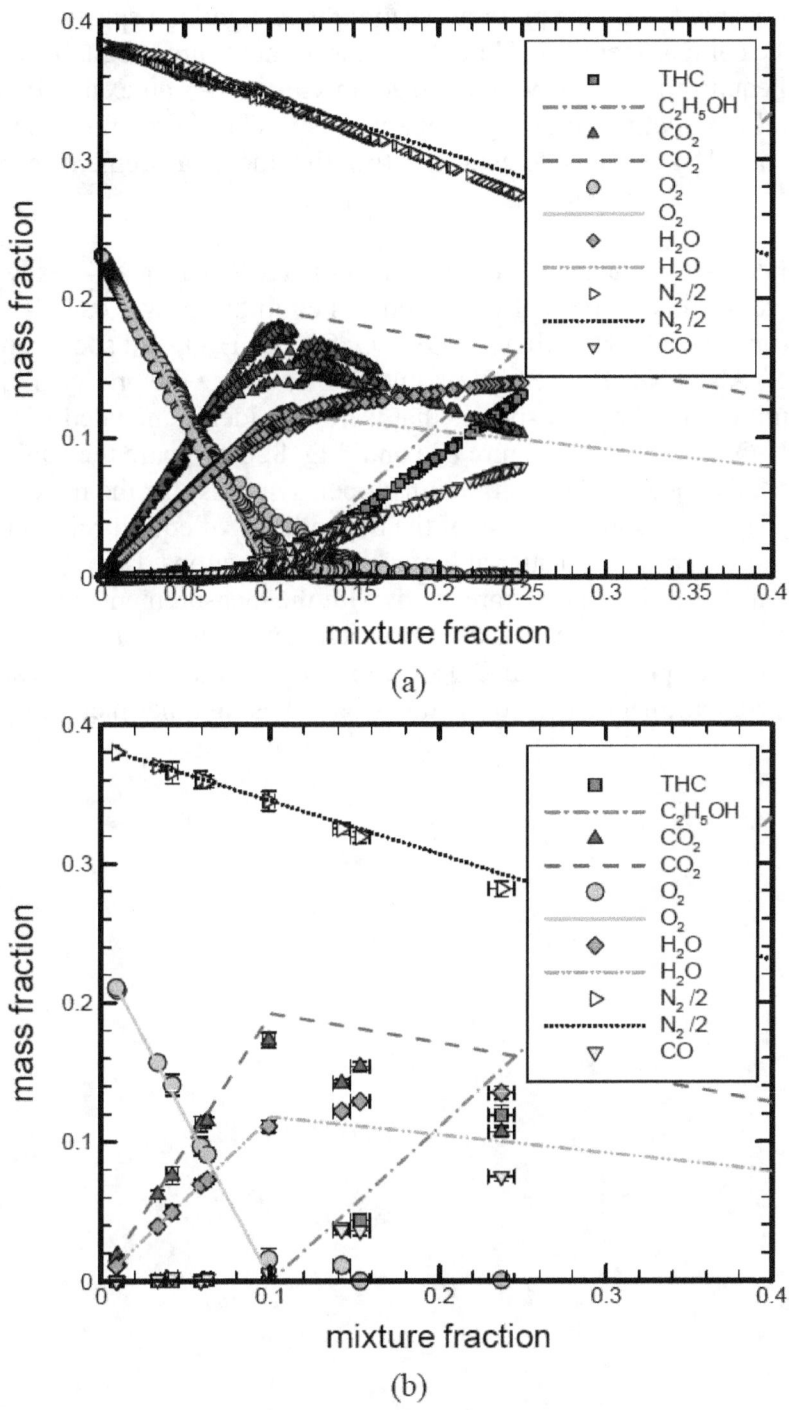

Figure 81 Mass fractions of front and rear compartment gas species as a function of mixture fraction for ethanol tests #9 and #11: (a) transient measurements and (b) time-averaged measurements during the period when the HRR was quasi-steady.

5.2 Chemical Equilibrium

This section is complementary to the previous section, investigating aspects of the species concentrations in the compartment fire. Here, the measurements are used to test the hypothesis that local chemical equilibrium exists within the compartment. As an example, results are presented for natural gas. Although this type of analysis has been previously considered (e.g., Sivathanu and Faeth, [41]), it is of interest to confirm that the measurements conducted in this study are consistent with previous results.

The time-averaged data taken during the period when the HRR was quasi-steady shown in Fig. 73 are considered in this analysis. Local chemical equilibrium was calculated using the measured species (unburned hydrocarbons, CO_2, CO, O_2, N_2, H_2O, and soot) and temperature data as input. The STANJAN software [45] was used to calculate the products of local thermodynamic equilibrium. Product species that were considered included CH_4, C_2H_6, C_3H_8, CO_2, CO, O_2, N_2, H_2O, soot and H_2. Figure 82a and Fig. 82b compare the equilibrium results (filled symbols) with the measured mass fractions (open symbols) for the major species, e.g. CH_4, CO_2, CO, O_2, H_2O, and soot. Because of the dependency of equilibrium on temperature, the calculated results show significant scatter at larger mixture fractions. Like the mixture fraction correlation, the calculations show good agreement with the measurements for fuel lean conditions, but significant differences exist for fuel rich conditions - consistent with previous results [41]. In particular, predictions of CO are entirely inaccurate using this approach. In the following sections, compartment chemistry is investigated using other methods.

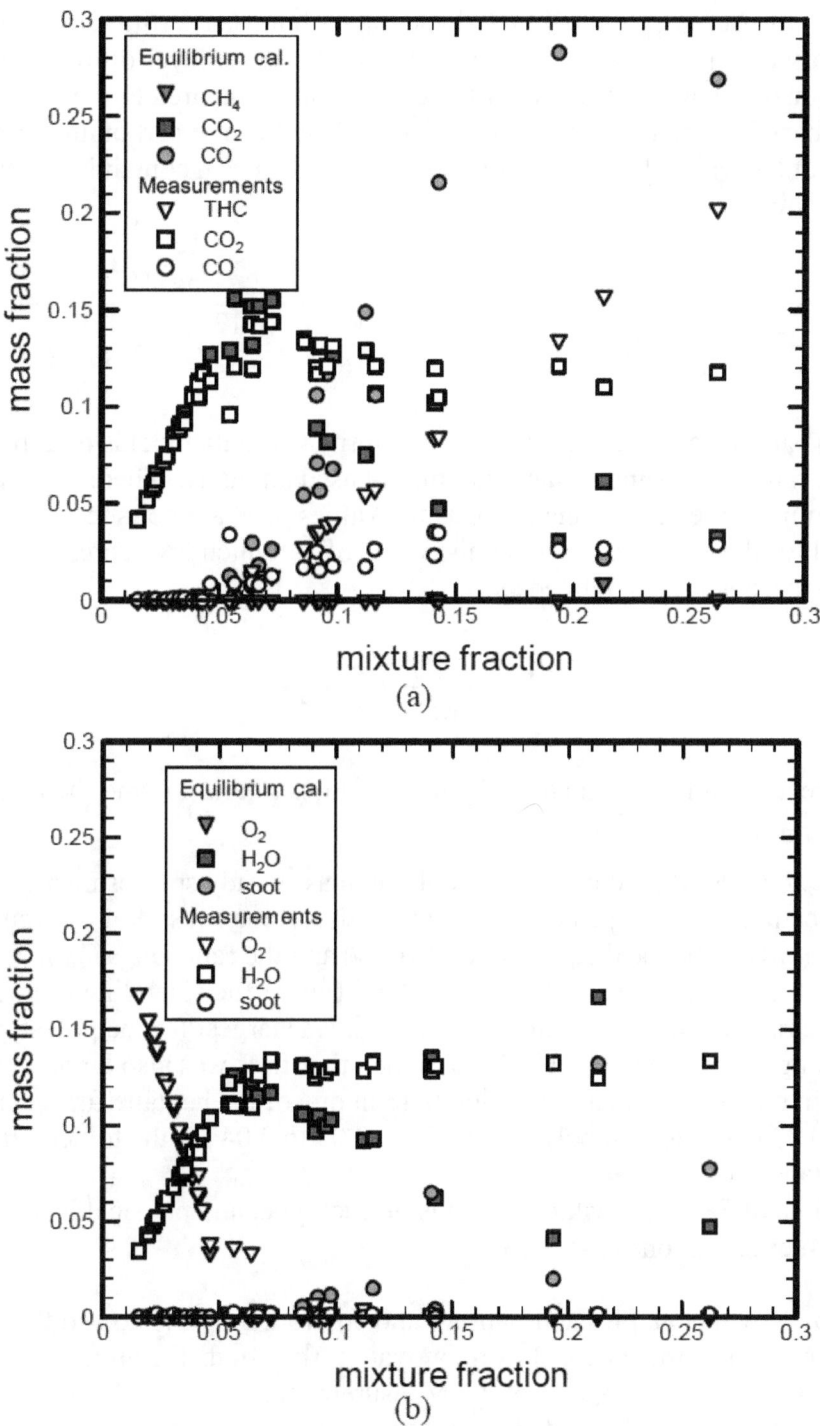

Figure 82 Comparison between the equilibrium calculations and the time-averaged species mass fractions measured in the front and rear of the compartment as a function of the mixture fraction during the period when the HRR was quasi-steady in the natural gas fire tests #1-#3, and #6.

5.3 Carbon Balance

The compartment measurements show that elemental carbon was partitioned among soot and three principal gaseous species (CO_2, CO, and CH_4) in the upper layer of the compartment. Other hydrocarbons were measured in only trace quantities compared to methane. The fractional mass-based amount of carbon that existed in the form of carbon monoxide (F_{CO}) or carbonaceous soot (F_{soot}) is related to the mass fractions of carbon containing species at each measurement location as:

$$F_{soot} = \frac{Y_{soot}}{\frac{12}{16}Y_{CH_4} + \frac{12}{44}Y_{CO_2} + \frac{12}{28}Y_{CO} + Y_{soot}} \; ; \quad F_{CO} = \frac{\frac{12}{28}Y_{CO}}{\frac{12}{16}Y_{CH_4} + \frac{12}{44}Y_{CO_2} + \frac{12}{28}Y_{CO} + Y_{soot}} \quad (28)$$

Typically, composition results are presented in the form of product yields or generation rates (defined below), rather than simply fractional mass-based amounts. There are some advantages, however, to examining the data in this form, as the values of F_i are bounded from 0 to 1. In the results presented for the compartment data, the value of X_s, which is a representation of the amount of carbonaceous soot is defined as:

$$X_s = \frac{Y_{soot}}{MW_c} \bigg/ \sum \frac{Y_i}{MW_i} \quad (29)$$

which comes directly from algebraic manipulation of Eqs. 16 and 17, and the facts that $\sum X_i MW_i$ is a constant, and $\sum X_i = 1$.

Table 23 lists F_{soot} and F_{CO} based on averages of the quasi-steady species measurements at the front and rear locations in the heptane, toluene and polystyrene fires. For convenience, the fire heat release rate (HRR), the local equivalence ratio (ϕ) and the ratio (F_{CO}/F_{soot}) are also included in the table. The value of F_{soot} was different for the different fuels, tending to increase with the local equivalence ratio (or mixture faction.) The F_{soot} was largest for the polystyrene fires (see Table 23), reaching a value of 0.66. The F_{soot} in the other fires was also large, taking on values as large as 0.29 in one of the toluene fires and 0.45 in one of the heptane fires. The value of F_{CO} was as large as 0.03 to 0.1 for the polystyrene fires, 0.03 to 0.04 for the toluene fires, and 0.02 to 0.14 for the heptane fires.

Table 23 lists value of F_{CO}/F_{soot}, which depends on fuel type, and physical location. Its value was less than 0.4, except in one case.

Measurements by Koylu et al. [46] and Santoro and co-workers [47] showed that there is a linear relation in the emission of soot and CO from buoyant turbulent diffusion flames burning various hydrocarbon fuels (acetylene, propene, etc.). Measurements in the fuel lean (overfire) plume region of hydrocarbon fires showed that the soot and CO generation factors (η_S and η_{CO}) tended to increase with flame residence time, until a near-constant value was reached after long times (compared to the smoke point). Koylu et al. [46] reported that the ratio of the CO and soot generation factors for a range of fuel types was such that, $\eta_{CO}/\eta_S = 0.34 \pm 0.09$. The generation rate was defined as the mass of soot (or gas species) produced per unit mass of fuel carbon consumed. This is slightly different than the soot (or gas species) yield (y_{CO} and y_S), which is based on the mass of all elements (not just carbon) in the fuel stream. The ratios of the yields and the generation rates, however, are equal, and their value can be determined at any location from the ratio of the mass fractions of CO and soot:

$$\eta_{CO}/\eta_S = y_{CO}/y_{soot} = Y_{CO}/Y_{soot} = (7/3)\, F_{CO}/F_{soot} \tag{30}$$

The constant value (7/3) in Eq. 30 is the ratio of the total CO mass to the mass of carbon.

Table 24 lists y_{CO}, y_{soot}, and the ratio y_{CO}/y_{soot} based on the time-averaged species measurements at the front and rear compartment locations and the stack when the heat release rate was quasi-steady during each of the fires (natural gas, methanol, ethanol, heptane, toluene and polystyrene). The fire heat release rate (HRR) and the local equivalence ratio (ϕ) are also listed. Much of the same data was used as in Table 23.

Figure 83a and Fig. 83b show the yield of CO and soot as a function of the local equivalence ratio for the heptane, toluene and polystyrene fires. Straight lines seem to be reasonable fits to the data, intersecting the intercept at a non-zero value. Figure 84 is analogous to Fig. 83, with the parameters F_{soot} and F_{CO} considered in lieu of y_{CO} and y_{soot}. The trends and values of the data shown in the graphs are very similar in appearance, consistent with the data presented in Table 23 and F_{soot} and F_{CO} is a reasonable way to represent the results and understand that in the toluene and polystyrene fires, almost twice as much carbon exists in the form of soot as compared to CO. The heptane data is much more scattered, making any sort of authoritative generality less convincing. In the richest polystyrene fire ($\phi \approx 2.5$), almost 80 % of the carbon exists in the form of CO or soot, with relatively little carbon in the form of CO_2.

Figure 85a and b shows the yield of CO and soot as a function of the local equivalence ratio for the natural gas and ethanol fires. The values of y_{CO} and y_{soot} were relatively low, as compared to the results for the smoky fuels presented in Fig. 83.

Figure 86 shows the CO yield as a function of the soot yield for the same quasi-steady data shown in Fig. 83 for the heptane, toluene and polystyrene fires. Also shown is a line representing the results of Koylu et al [46]. Koylu reported about 30 % scatter in the ratio of the yields of CO to soot, which is considerably smaller than that seen in the figure. Nevertheless, more data are needed to examine this relationship in the upper layer of compartment fires. It is interesting to note that Tewarson et al [48] reported that the ratio of the CO and soot generation efficiencies from small fires burning polymers varied, depending on the exact fuel type and the amount of ventilation.

Figure 87 shows the ratio of the CO yield to the soot yield as a function of the local equivalence ratio for the same quasi-steady data shown in Fig. 83 and Fig. 86 for the heptane, toluene and polystyrene fires. Best fit lines for each fuel type are different, and highlight the trends in the y_{CO}/y_{soot} data.

Figure 88 shows the CO yield as a function of the soot yield for the same quasi-steady data shown in Fig. 85 for the natural gas and ethanol fires. Results for methanol are not shown, because the measured soot yield was zero. Also shown in the figure is a line representing the results reported by Koylu et al [46] for rather heavy hydrocarbons, including propane, and propylene. The Koylu results do not agree with the current set of data for these non-smoky fuels. Figure 89 shows the ratio of the CO yield to the soot yield as a function of the local equivalence ratio for the same quasi-steady data shown in Fig. 85 for the natural gas and ethanol fires. Both sets of data could be fit by straight lines. The figure shows that a horizontal line could adequately represent the natural gas results for rich conditions ($\phi>1$). The ethanol results, however, show a finite slope, which is unique among the fuels tested (also see Fig. 87). Experiments over a range of ventilation conditions in full-scale are planned to further investigate the consistency and repeatability of the trends in the CO and soot measurement results.

Table 23. Average fractional soot, CO and CO/Soot ratio at the front and rear compartment measurement locations.

Fuel	HRR	Rear				Front			
		ϕ_{local}	F_{CO}	F_{soot}	F_{CO}/F_{soot}	ϕ_{local}	F_{CO}	F_{soot}	F_{CO}/F_{soot}
heptane	140	0.957	0.047	0.049	0.959	1.463	0.019	0.039	0.49
	148	0.563	0.01	0.035	0.280	0.581	0.015	0.041	0.36
	153	0.686	0.016	0.031	0.532	0.660	0.013	0.034	0.37
	160	0.547	0.013	0.066	0.205	0.677	0.022	0.058	0.38
	221	1.214	0.321	0.168	1.911	2.024	0.078	0.263	0.30
	227	0.669	0.039	0.094	0.421	1.080	0.101	0.142	0.72
	246	0.650	0.017	0.064	0.267	1.169	0.076	0.116	0.65
	269	0.730	0.012	0.043	0.288	1.922	0.161	0.089	1.8
	301	1.366	0.103	0.170	0.603	2.407	0.22	0.195	1.13
	341	1.249	0.143	0.211	0.678	1.636	0.22	0.236	0.93
	375	1.163	0.101	0.162	0.622	1.855	0.187	0.257	0.73
	377	2.281	0.138	0.238	0.578	3.469	0.216	0.267	0.81
toluene	138	0.540	0.044	0.190	0.230	0.741	0.04	0.195	0.21
	202	0.599	0.009	0.117	0.077	0.981	0.022	0.179	0.13
	295	0.989	0.026	0.206	0.127	1.244	0.061	0.291	0.21
polystyrene	15	0.137	0.074	0.149	0.498	0.133	0.071	0.157	0.45
	67	0.394	0.075	0.216	0.347	0.374	0.071	0.205	0.35
	309	2.017	0.195	0.583	0.335	1.307	0.191	0.337	0.57
	358	2.391	0.231	0.650	0.355	1.884	0.234	0.555	0.42

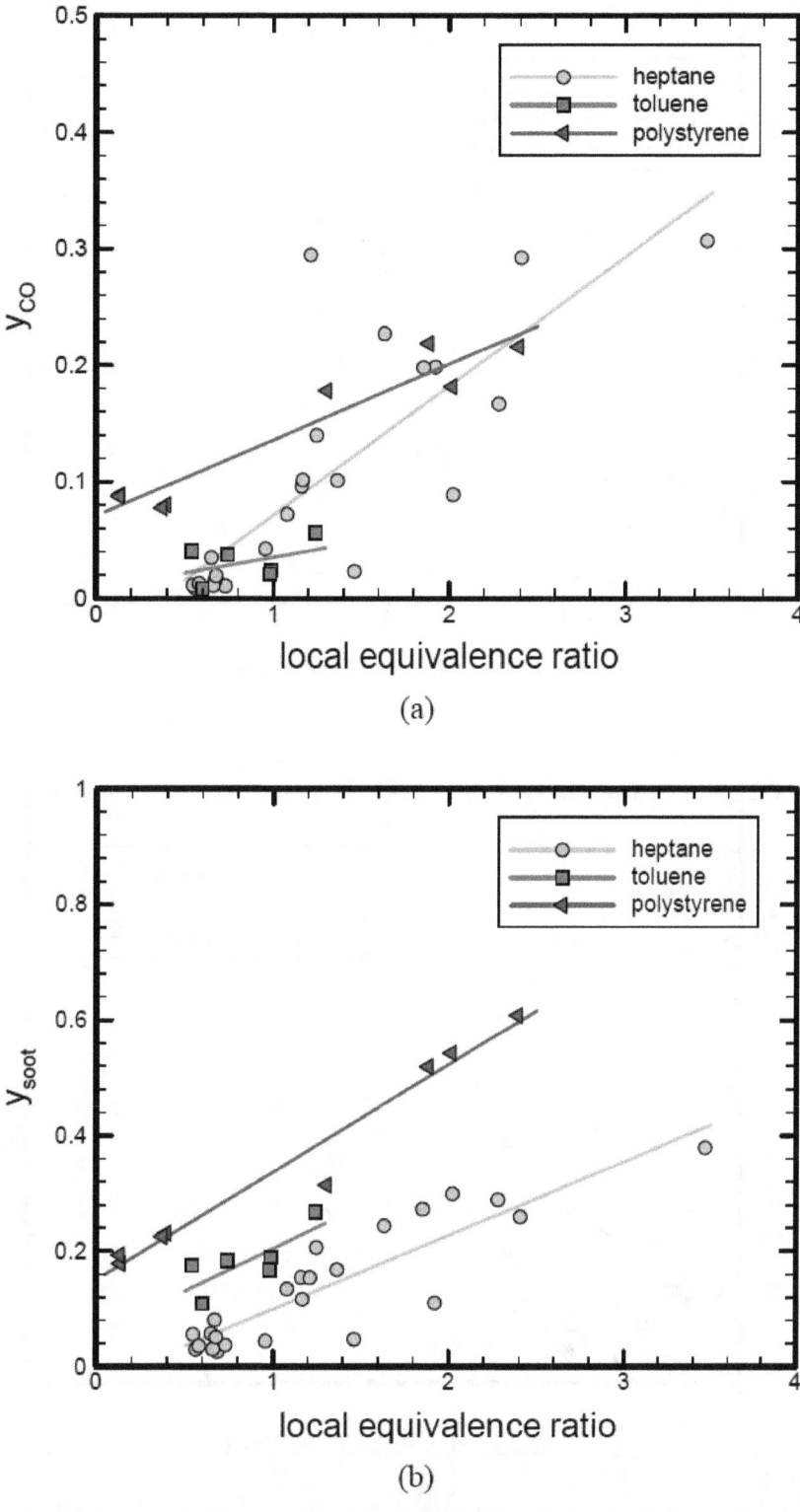

Figure 83 The CO and soot yields as a function of the local equivalence ratio for the time-averaged measurements during the period when the HRR was quasi-steady in the heptane, toluene and polystyrene fires.

105

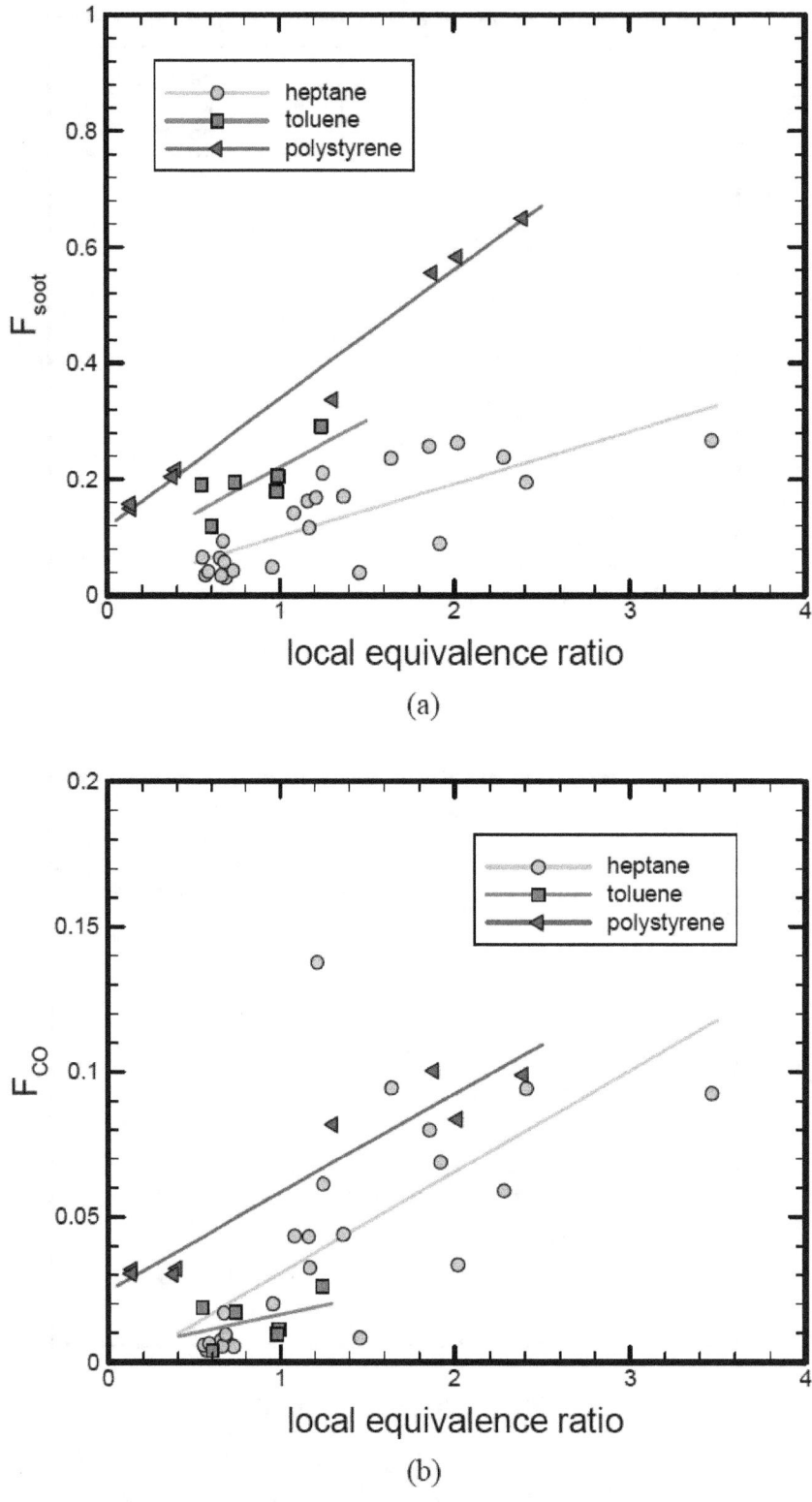

Figure 84. The values of F_{CO} and F_{Soot} as a function of the local equivalence ratio during the period when the HRR was quasi-steady in the heptane, toluene and polystyrene fires.

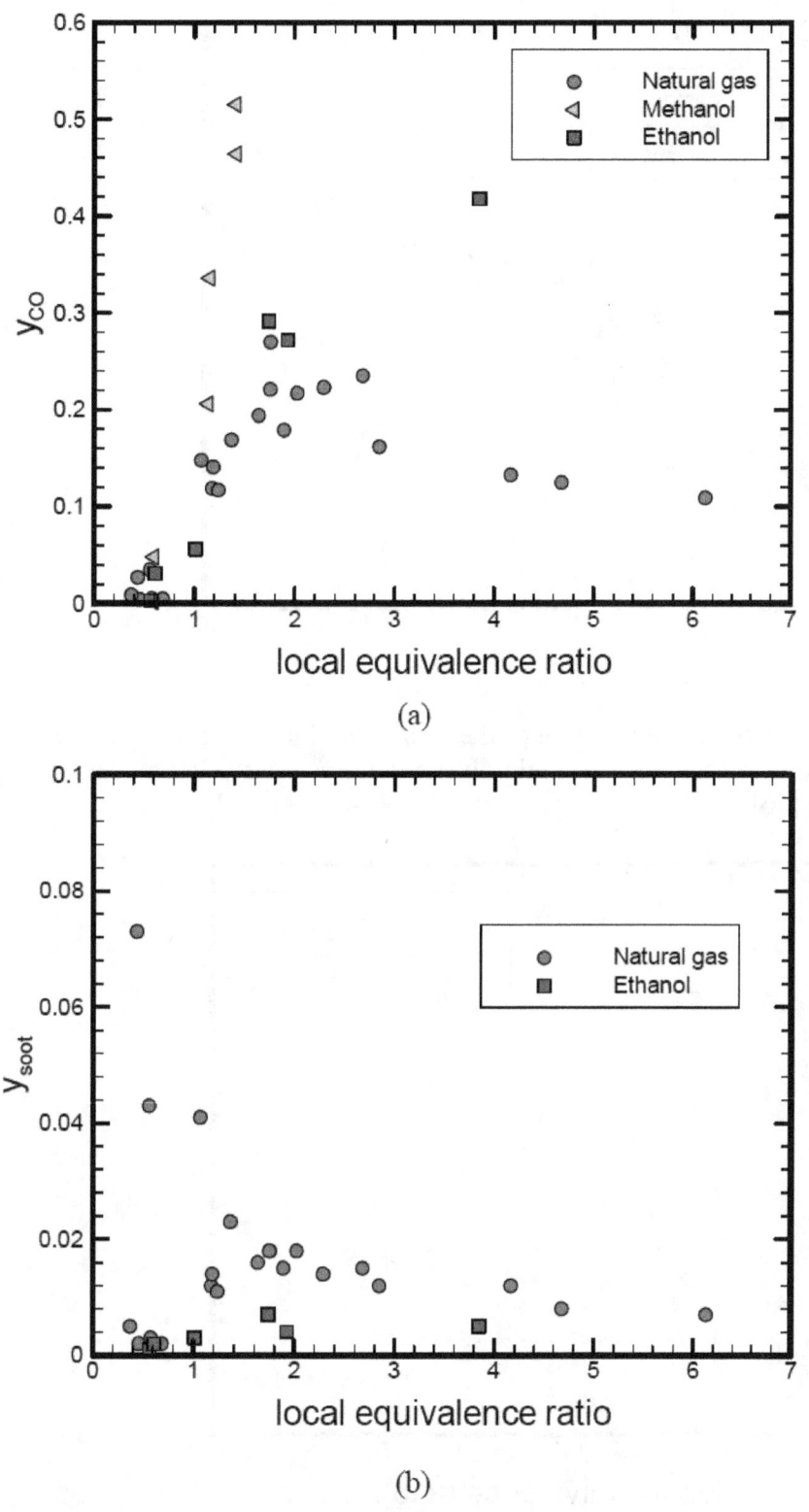

Figure 85 The CO and soot yields as a function of the local equivalence ratio during the period when the HRR was quasi-steady in the natural gas and ethanol fires.

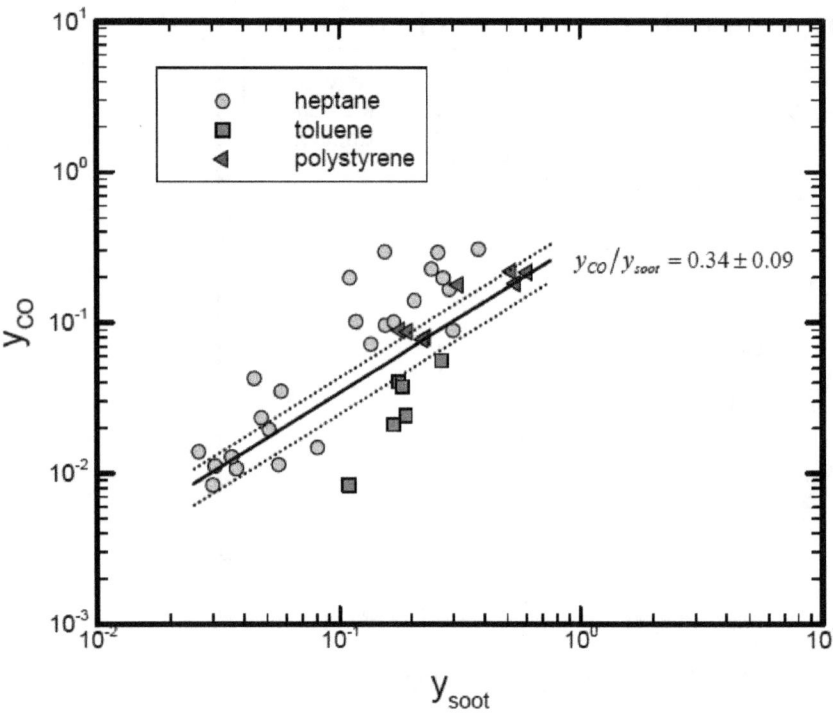

Figure 86. The CO yield as a function of the soot yield during the period when the HRR was quasi-steady in the heptane, toluene and polystyrene fires. Also shown is a line representing the results of Koylu [46].

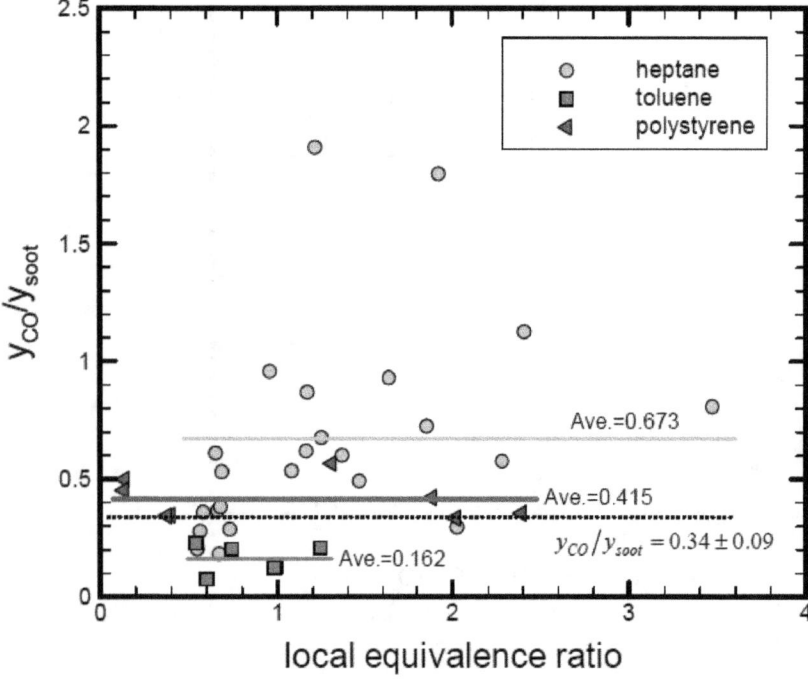

Figure 87. The ratio of the CO to soot yield as a function of the local equivalence ration during the period when the HRR was quasi-steady in the heptane, toluene and polystyrene fires. Best fit lines to the data and a line representing the results of Koylu [46] are also shown.

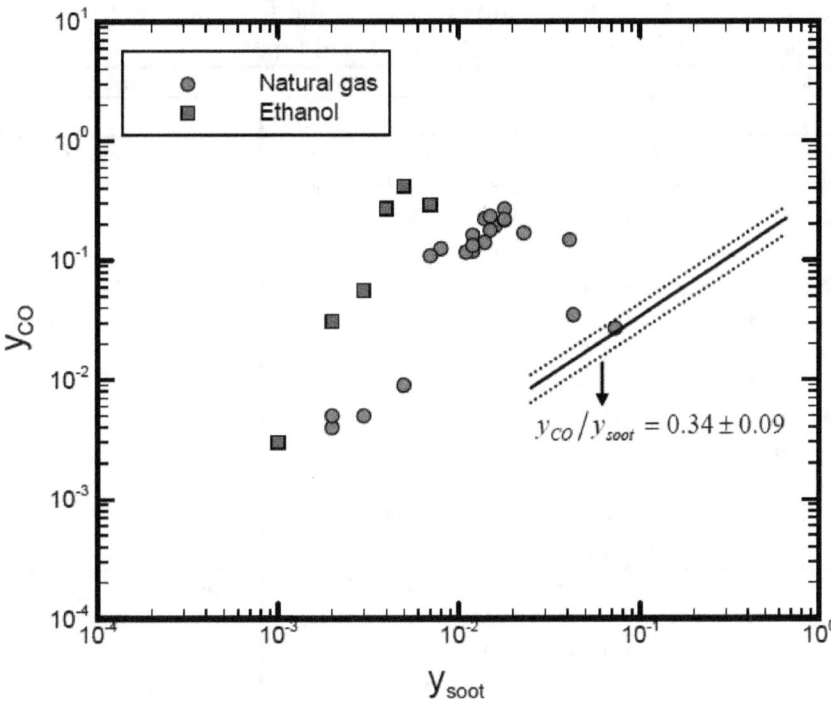

Figure 88 The CO yield as a function of the soot yield for the same data shown in Fig. 85 during the period when the HRR was quasi-steady in the natural gas and ethanol fires.

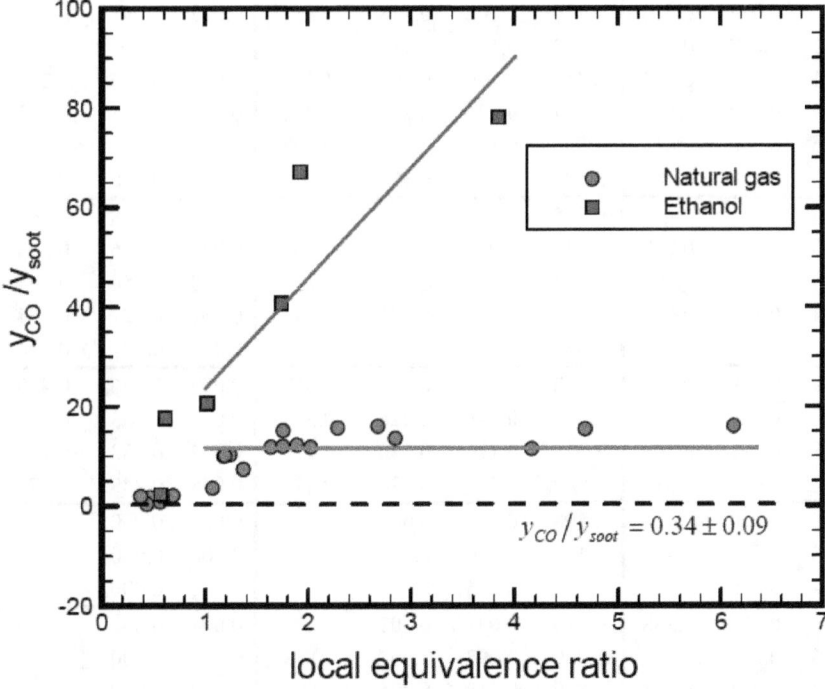

Figure 89 The ratio of the CO to soot yield as a function of the local equivalence ration for the same data shown in Fig. 85 during the period when the HRR was quasi-steady in the natural gas and ethanol fires.

Table 24. Time-averaged yields of soot, CO, and the ratio (y_{CO}/y_s) at the front and rear compartment measurement locations and in the exhaust stack for all fuel types.

Fuel	HRR [kW]	Rear				Front				Stack		
		φ_{local}	y_{co}	y_s	y_{co}/y_s	φ_{local}	y_{co}	y_s	y_{co}/y_s	y_{co}	y_s	y_{co}/y_s
Natural gas	74	0.686	0.005	0.002	2.126	0.579	0.005	0.003	1.933	-	-	-
	74	0.460	0.004	0.002	1.802	0.373	0.009	0.005	1.955	-	-	-
	75	0.559	0.035	0.043	0.817	0.436	0.027	0.073	0.373	-	-	-
	85	0.452	0.026	-	-	0.407	0.015	-	-	0.067	0.000	-
	97	0.565	0.058	-	-	0.481	0.036	-	-	0.161	0.000	-
	179	1.239	0.117	0.011	10.35	1.188	0.141	0.014	10.14	-	-	-
	181	0.980	0.068	-	-	1.070	0.192	-	-	0.018	0.000	
	186	1.368	0.169	0.023	7.396	1.069	0.148	0.041	3.588	-	-	
	265	1.891	0.179	0.015	12.30	1.753	0.221	0.018	12.01	-	-	-
	272	2.849	0.162	0.012	13.58	4.164	0.133	0.012	11.51	-	-	-
	273	1.372	0.191	-	-	1.720	0.279	-	-	0.016	0.000	-
	417	4.681	0.125	0.008	15.45	6.125	0.109	0.007	16.07	-	-	-
	424	2.025	0.217	0.018	11.86	2.679	0.235	0.015	16.03	0.024	0.000	
Heptane	83	0.383	0.015	-	-	0.349	0.022	-	-	0.009	0.044	0.212
	88	0.394	0.007	-	-	0.347	0.005	-	-	0.006	0.029	0.212
	140	0.957	0.043	0.044	0.957	1.463	0.023	0.047	0.491	-	-	-
	148	0.563	0.008	0.030	0.280	0.581	0.013	0.036	0.359	0.007	0.030	0.234
	153	0.686	0.014	0.026	0.531	0.660	0.011	0.031	0.366	-	-	-
	160	0.547	0.011	0.056	0.204	0.677	0.020	0.051	0.382	0.010	0.056	0.174
	221	1.216	0.294	0.156	1.887	2.038	0.088	0.305	0.289	-	-	-
	227	0.670	0.015	0.081	0.183	1.081	0.072	0.136	0.531	0.021	0.118	0.176
	246	0.651	0.035	0.057	0.610	1.170	0.102	0.118	0.864	0.022	0.095	0.237
	269	0.730	0.011	0.037	0.287	1.923	0.198	0.111	1.783	-	-	-
	301	1.369	0.101	0.170	0.593	2.417	0.290	0.264	1.100	0.052	0.230	0.228
	341	1.253	0.139	0.209	0.667	1.644	0.226	0.248	0.912	0.051	0.197	0.257
	375	1.165	0.096	0.156	0.613	1.866	0.196	0.277	0.708	-	-	-
	377	2.294	0.165	0.294	0.562	3.508	0.300	0.389	0.772	0.064	0.240	0.265
Toluene	49	0.312	0.115	-	-	0.275	0.124	-	-	0.098	0.148	0.662
	138	0.541	0.040	0.176	0.229	0.742	0.037	0.185	0.203	0.045	0.146	0.305
	202	0.600	0.008	0.110	0.076	0.983	0.021	0.169	0.124	0.013	0.154	0.084
	295	0.991	0.024	0.191	0.125	1.252	0.056	0.273	0.204	0.011	0.127	0.084
	339	0.861	0.104	-	-	0.936	0.200	-	-	0.023	0.171	0.136
Polystyrene	15	0.137	0.089	0.179	0.497	0.133	0.087	0.194	0.452	0.072	0.176	0.410
	67	0.395	0.080	0.232	0.345	0.374	0.078	0.226	0.343	0.070	0.253	0.277
	309	2.117	0.173	0.561	0.308	1.319	0.176	0.320	0.550	0.017	0.133	0.128
	358	2.572	0.200	0.631	0.317	1.962	0.210	0.536	0.391	0.044	0.249	0.178
Methanol	143	0.592	0.002	0.000	-	0.598	0.048	0.000	-	0.000	0.000	-
	240	1.137	0.206	0.000	-	1.151	0.336	0.000	-	0.006	0.000	-
	306	1.414	0.464	0.000	-	1.411	0.515	0.000	-	0.011	0.000	-
Ethanol	144	0.565	0.003	0.001	2.288	0.610	0.031	0.002	17.71	0.004	0.000	-
	263	1.014	0.056	0.003	20.68	1.739	0.292	0.007	40.68	0.004	0.000	-
	335	1.927	0.272	0.004	67.13	3.850	0.418	0.005	78.05	0.008	0.000	-

5.4 Post-Compartment Product Yields

It is useful to consider the product yields downstream from the fire compartment in the exhaust stack, and to compare these results to conditions in the compartment. These comparisons highlight the effects of a compartment on fire chemistry.

The yields of CO_2, CO, soot, and hydrocarbons determined from the measurements made in the exhaust hood during the quasi-steady burning periods for each of the fuels tested are shown in Figure 90 through Fig. 93 as a function of the fire heat release rate. The largest yield in the stack was for CO_2 (125 % to 300 %), followed by soot (< 25 %), CO (< 10 %), and total hydrocarbons (< 3.5 %). The CO_2 yield in the stack shown in Fig. 90 was related to the stoichiometry (Eq. 9) and the combustion efficiency. The CO yield in the stack shown in Fig. 91 appeared to be a function of fuel type and fire size. Some fuels exhibited high CO yields for lower heat release rates (toluene, polystyrene, ethanol, natural gas) and other fuels exhibited relatively higher CO yields (heptane, methanol) for higher heat release rates. The results for the soot yield shown in Fig. 92 were also dependent on the fuel type and the HRR, whereas some fuels (natural gas and methanol) produced absolutely negligible amounts of soot. The soot yield was as large as 15 % to 25 % for the heptane, toluene and polystyrene fires.

The local product species yields in the rear and front of the compartment were discussed previously. Table 24 lists the CO and soot yields in the smoke-laden (heptane, toluene and polystyrene) and non-smoky fires (natural gas and ethanol) as a function of the fire size, both in the compartment and in the stack. The data in the Table are also plotted in Fig. 83 and Fig. 85. For the smoky fires, the yield of soot was almost always larger in the compartment than in the stack. For example, the yield of soot in the toluene fire varied from 0.15 to 0.17 in the stack as compared to values as high as 0.27 in the compartment. The polystyrene fire was the smokiest and the yield of soot varied from 0.13 to 0.25 in the stack as compared to 0.18 to 0.61 in the compartment. The yield of CO was generally, but not always larger in the compartment than in the stack, and typically a factor of 1.5 to 7 times larger depending on the fuel type. The yield of CO in the polystyrene fire varied from 0.02 to 0.07 in the stack as compared to 0.08 to 0.26 in the compartment. The yield of CO in the ethanol fire varied from approximately 0.004 to 0.07 in the stack as compared to 0.003 to 0.42 in the compartment.

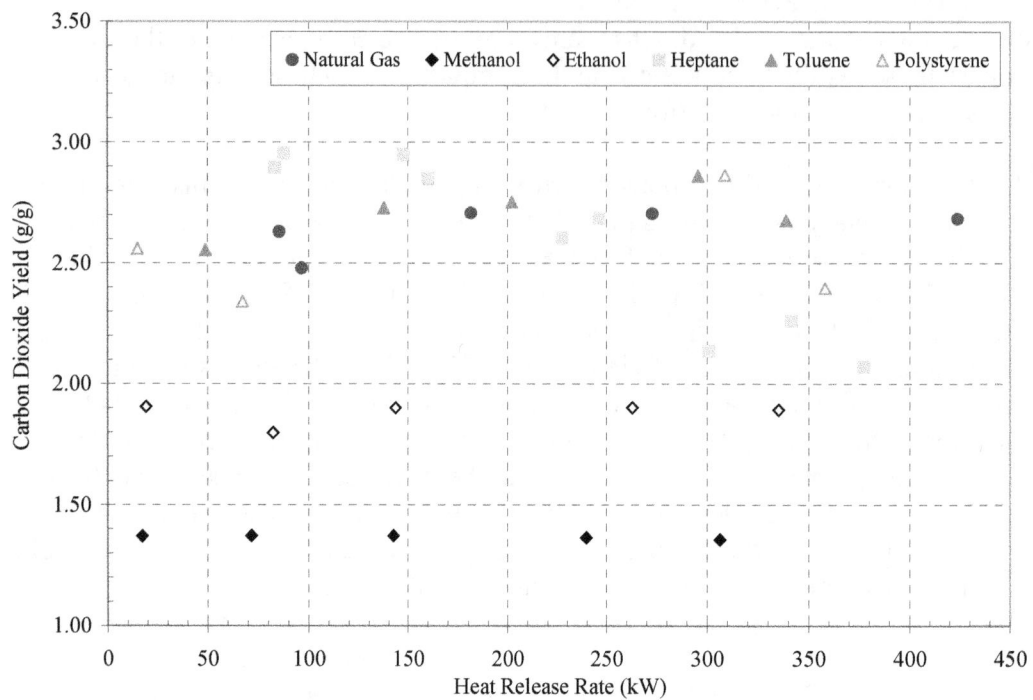

Figure 90 The CO_2 yield in the exhaust stack as a function of the fire heat release rate during the periods when the HRR was quasi-steady for each of the fuels tested.

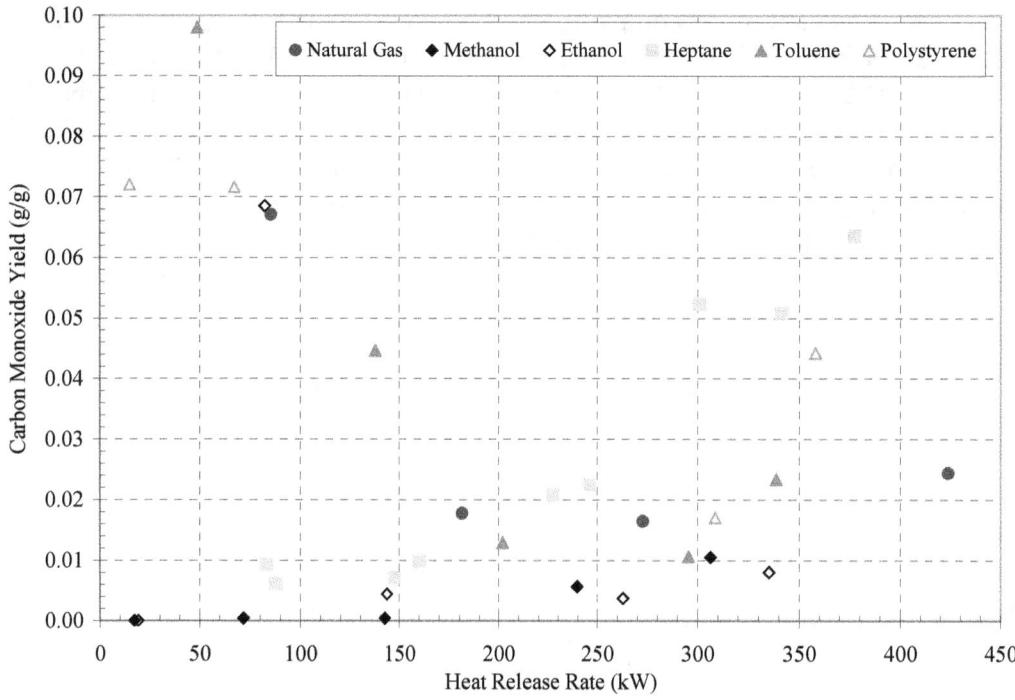

Figure 91. The CO yield in the exhaust stack as a function of the fire heat release rate during the periods when the HRR was quasi-steady for each of the fuels tested.

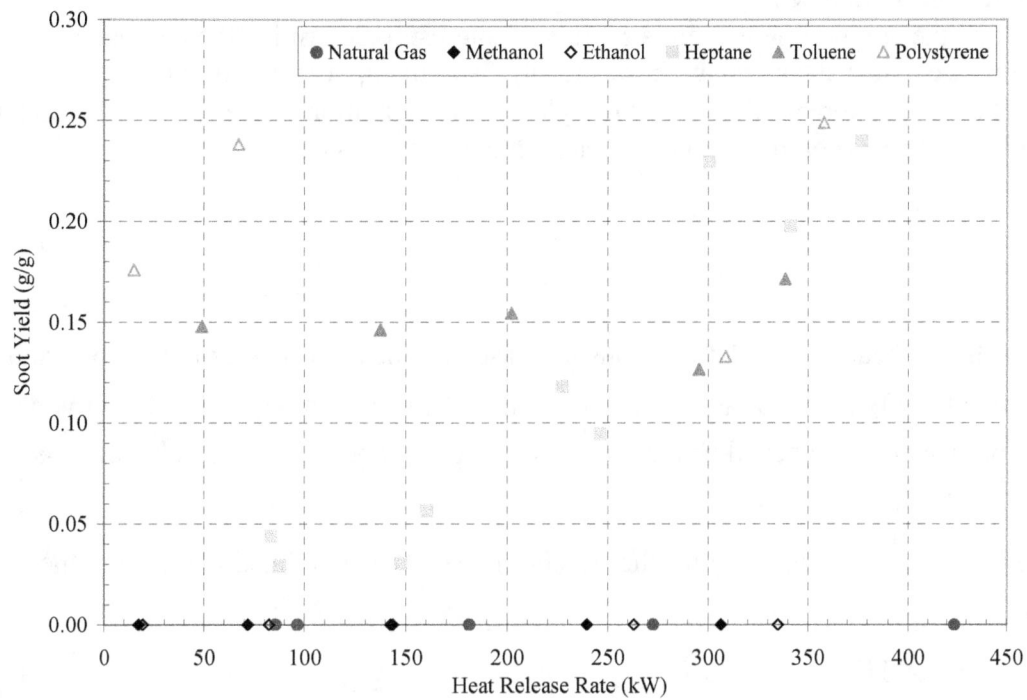

Figure 92. The soot yield in the exhaust stack as a function of the fire heat release rate during the periods when the HRR was quasi-steady for each of the fuels tested.

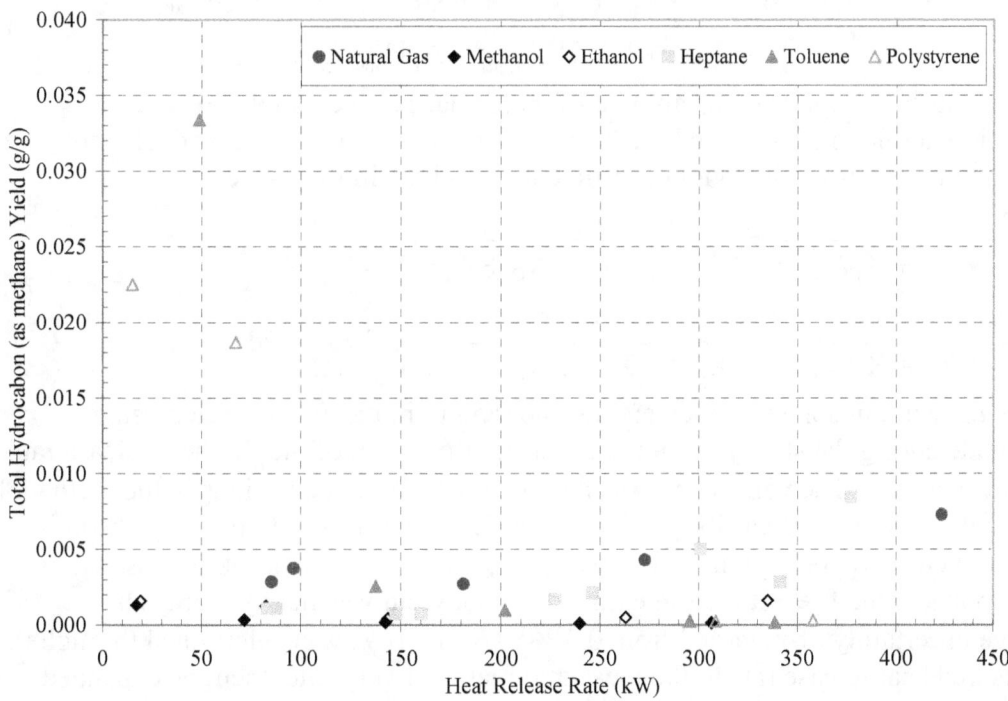

Figure 93. The total hydrocarbon yield in the exhaust stack as a function of the fire heat release rate during the periods when the HRR was quasi-steady for each of the fuels tested.

113

5.5 Combustion Efficiency

To better understand the compartment chemistry, it is of interest to determine the combustion efficiency both in the exhaust stack and at various locations in the upper layer of the compartment. The combustion efficiency (χ_a) is a global representation of the fractional amount of heat released by the fire as compared to complete combustion. It is defined as:

$$\chi_a = \frac{\Delta H_c}{\Delta H_{c,ideal}} \tag{31}$$

where $\Delta H_{c,ideal}$ is the net heat of complete combustion based on the conversion of all carbon and hydrogen in the fuel to CO_2 and H_2O (assumed to remain in the vapor phase) and ΔH_c is the net heat of combustion, which is the actual heat released in a chemical reaction. The value of χ_a is bounded by 0 % and 100 %.

Using the nomenclature defined in Eq. 9 for the stoichiometry of the combustion reaction, the value of $\Delta H_{c,ideal}$ is:

$$\Delta H_{c,ideal} = x\Delta_f H^\circ_{CO_2,gas} + \frac{y}{2}\Delta_f H^\circ_{H_2O,gas} - \Delta_f H^\circ_{fuel,(gas,liquid,solid)} \tag{32}$$

where $\Delta_f H^\circ_{i,state}$ is the heat of formation of species i at a given state. The heats of formation of CO_2 and H_2O are given in Ref. [49]. The value of ΔH_c, the net heat of combustion is given by:

$$\Delta H_c = b\,\Delta_f H^\circ_{CO_2,gas} + c\,\Delta_f H^\circ_{CO,gas} + e\,\Delta_f H^\circ_{C,solid} + d\,\Delta_f H^\circ_{CH_4,gas}$$
$$+ f\,\Delta_f H^\circ_{H_2O,gas} - \Delta_f H^\circ_{fuel,(gas,liquid,solid)} \tag{33}$$

where the coefficients b – f represent the amount of molecular products in the general combustion reaction defined in Eq. 9, and the heats of formation of CO, soot, and CH_4 are given by Ref. [49] . The molecular product yield coefficients are given in Eq. 34 below.

$$b = \frac{X_{CO_2}\,x}{\left(X_{CO_2} + X_{CO} + X_C + X_{CH_4}\right)}, c = \frac{X_{CO}\,x}{\left(X_{CO_2} + X_{CO} + X_C + X_{CH_4}\right)}$$

$$e = \frac{X_C\,x}{\left(X_{CO_2} + X_{CO} + X_C + X_{CH_4}\right)}, d = \frac{X_{CH_4}\,x}{\left(X_{CO_2} + X_{CO} + X_C + X_{CH_4}\right)}, f = \frac{y}{2} - 2d \tag{34}$$

Figure 94 shows the combustion efficiency and its uncertainty in the exhaust stack using measurements made during the steady burning periods as a function of the fire heat release rate for each of the fuels tested. The value of χ_a was nearly 100 % for all conditions in the methanol, ethanol and natural gas fires, whereas its value was smaller in the fires with the largest soot yields (heptane, toluene and polystyrene). The value of χ_a in the exhaust stack was as low as 75 % during the polystyrene fire. At the same time, the soot yield was nearly 0.25. The expanded relative uncertainty of χ_a varied from 0.3 % to 6 %. If χ_a was determined through the ratio of the measured heat release rate to the measured mass delivery rate, then the expanded relative uncertainty was larger than 15 % and was not a function of the value of χ_a.

Figure 95 shows the combustion efficiency in the rear and front compartment sampling locations as a function of the measured fire heat release rate during the steady burning periods for the three smokiest fuels (heptane, toluene and polystyrene). There was a large amount of scatter in the results. In general, the value of χ_a tended to decrease with increasing values of the HRR. During one of the larger polystyrene fires, the value of χ_a was as low as 45 % in the compartment. For the same fire, the value of χ_a was much larger in the exhaust stack, reaching almost 80 %. The value of χ_a for the heptane fires varied between 50 % and 100 %. The scatter in the heptane results was rather large. As expected, the value of χ_a tended to be higher when the oxygen volume fraction was larger as seen in Fig. 96. A comparison of Fig. 94 and Fig. 95 shows that the value of χ_a inside the compartment was typically, but not always, equal to or smaller than the value of χ_a in the stack, consistent with the idea that incomplete products of combustion continue to oxidize once they exit the compartment and are exposed to air.

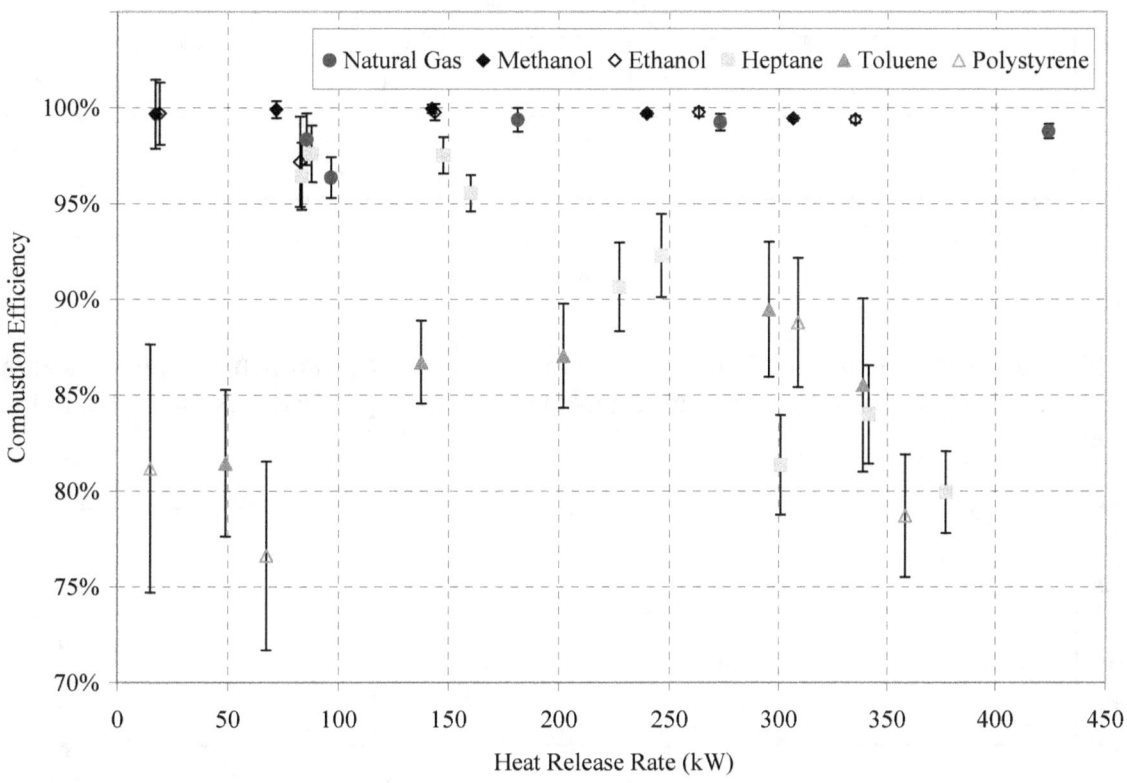

Figure 94. The combustion efficiency in the exhaust stack as a function of the fire HRR.

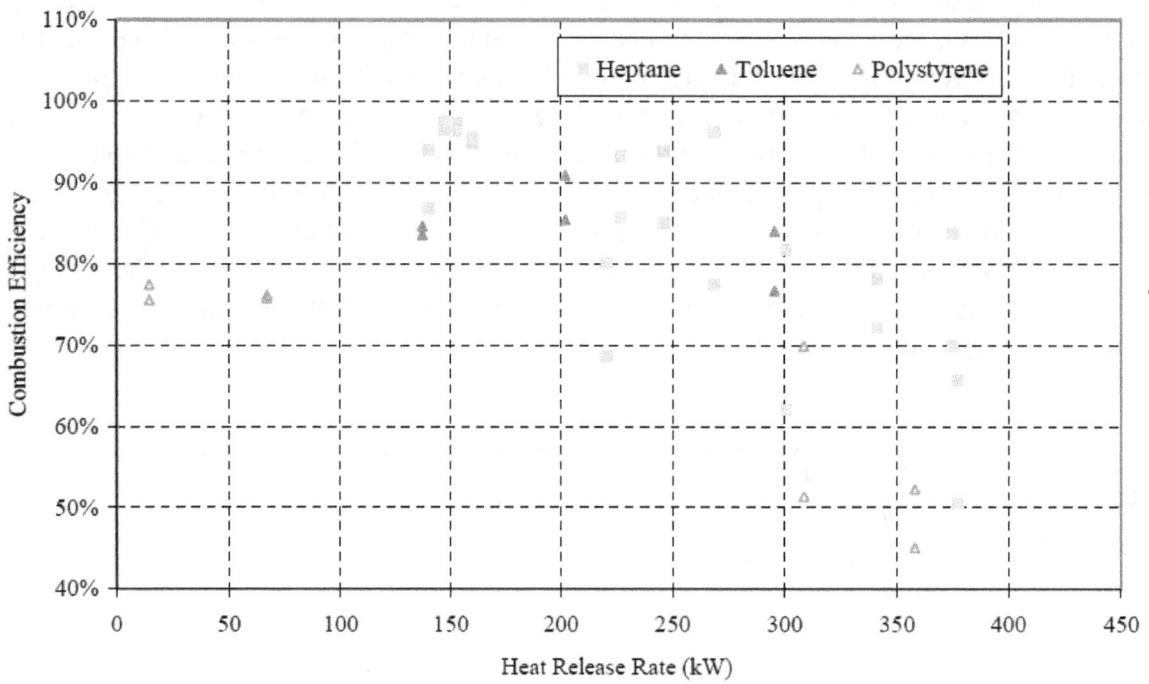

Figure 95. The local combustion efficiency at the rear and front compartment sampling locations as a function of the fire heat release rate during the periods when the HRR was quasi-steady for three fuels.

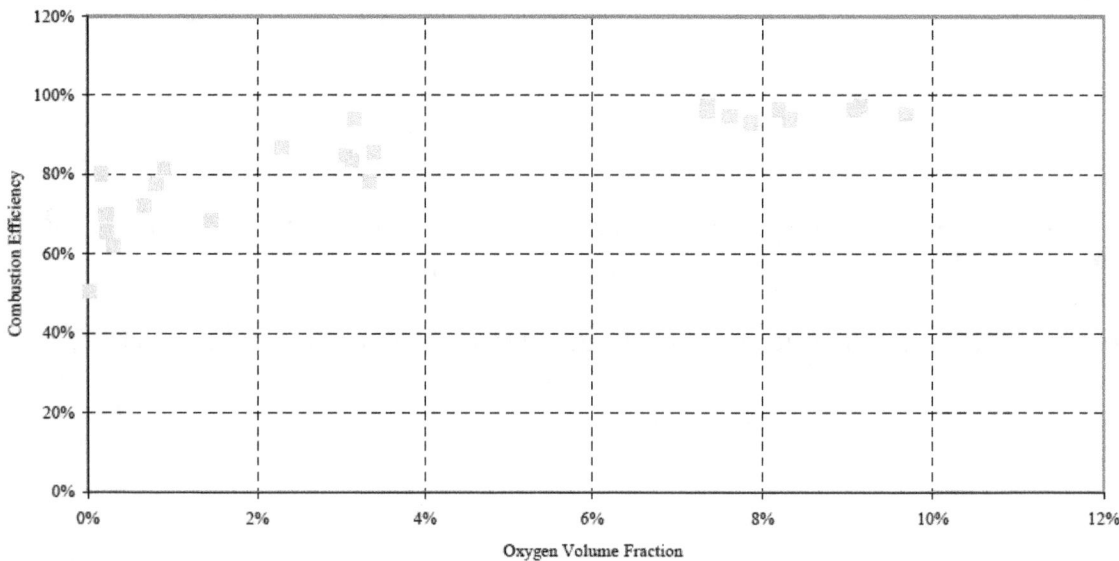

Figure 96. The local combustion efficiency at the rear and front compartment sampling locations as a function of the oxygen volume fraction during the periods when the HRR was quasi-steady in the heptane fire.

116

5.6 Summary of Chemical Analysis

The mass fractions of gas species, such as hydrocarbons, carbon dioxide, carbon monoxide, oxygen, water vapor, and nitrogen, were considered as a function of the mixture fraction for each fuel type. The measurements were compared with the state relations based on the mixture fraction correlation model. Measurements in the upper layer of the compartment showed that the CO volume fraction was a function of fuel type, fire size, ventilation opening and specific sample location. In addition, the gas chromatographic results showed that hydrocarbons in the upper layer were composed mainly of CH_4.

Consistent with previous studies, plotting the local composition as a function of the mixture fraction (or equivalence ratio) collapses hundreds of individual species measurements from an assortment of compartment conditions, with varying heat release rates, ventilation openings, and spatial locations, into a few coherent lines or bands. The results show that CO is not well-correlated with mixture fraction. Recognizing that CO kinetics are relatively slow, it was shown that consideration of temperature effects through analyses based on complete local thermal equilibrium and modification of the mixture fraction model to account for direct temperature correlation were not particularly fruitful. That these simple methods were not successful does not mean that improved accuracy cannot be achieved.

The results presented here demonstrate that it is important to consider soot as part of the mixture fraction analysis of compartment fires. This is particularly true in the upper layer of smoky fires in which about half (or more) of the fuel carbon may exist in the form of carbonaceous soot. Inclusion of soot in the analysis allows identification of fuel rich or underventilated conditions, conditions that otherwise would be considered lean or overventilated.

The combined mass of carbon in the form of soot and total hydrocarbons seem to be correlated by the soot-based mixture fraction for the conditions tested in this study. The CO and soot yields were found to be a function of fuel type and local equivalence ratio in the upper layer of the compartment for the smoky heptane, toluene and polystyrene fires. In addition, the ratio of CO yield to soot yield was found to be independent of local equivalence ratio for the smoky fuels. Additional experiments are needed to test this finding for other conditions and for full-scale compartment fires. In summary, mixture fraction is a useful initial way to describe the chemical structure of compartment fires.

6 Scaling Discussion

The applicability of the experimental results reported here to other compartment fire scenarios can be considered in terms of a number of normalized parameters traditionally used in fire modeling applications. Use of normalized parameters facilitates comparison of results from scenarios of different scales by normalizing key physical characteristics of the scenario. A number of different forms of scaling may be considered, depending on the fire phenomena of interest [50]. Table 25 lists three normalized parameters that may be used to compare fire scenarios with the experiments reported here. The ranges of values for the normalized parameters examined in this study are listed in the table. The table is intended to provide guidance when evaluating the applicability of the data set reported here. For any given fire scenario, more than one normalized parameter may be necessary for determining applicability of the validation results, depending on the parameters of interest. In this sense, the Table should be considered illustrative, not exhaustive.

Table 25. List of non-dimensional scaling parameters for compartment fires and the range of values examined in this study.

Parameter	Normalized Representation	Range of Values	Fuel Type
Heat Release Rate	$Q_d^* = \dfrac{\dot{Q}}{\rho_\infty c_p T_\infty \sqrt{g D D^2}}$	1.6 to 44.2 0.7 to 8.0 1.0 to 7.2 1.6 to 5.5 1.5 to 5.5 1.6 to 5.0	Natural Gas Heptane Toluene Polystyrene Methanol Ethanol
Ventilation	$\phi = \dfrac{\dot{m}_F / \dot{m}_{O_2}}{r}$, where $\dot{m}_{O_2} = \dfrac{0.23}{2} A_o \sqrt{h_o}$	0.10 to 1.71 0.16 to 0.83 0.09 to 0.96 0.03 to 0.69 0.04 to 0.75 0.01 to 0.77	Natural Gas Heptane Toluene Polystyrene Methanol Ethanol
Compartment height	$\dfrac{H}{D^*}$, where $D^* = \left(\dfrac{\dot{Q}}{\rho_\infty c_p T_\infty \sqrt{g}} \right)^{2/5}$	1.4 to 3.5 1.5 to 2.7 1.6 to 3.4 1.6 to 5.2 1.6 to 5.0	Natural Gas Heptane Toluene Polystyrene Methanol Ethanol

The most important parameter of any fire experiment is the heat release rate, as its magnitude drives changes in the thermal environment of the compartment or space of interest. A normalized quantity that relates the heat release rate to the diameter of the fire, D, is the first entry in Table 25, commonly known as Q_d^*, where $\dot{Q}$ is the heat release rate (kW), ρ_∞ is the ambient density (kg/m³), T_∞ is the ambient temperature (K), c_p is the specific heat (kJ/kg-K), and g is the acceleration of gravity (m/s²). A large value of Q_d^* represents a fire with a relatively large value of energy output power compared to its physical diameter, like an oil well blowout fire. A low value of Q_d^* represents a fire with a relatively small value of energy output compared to its diameter, like a smoldering fire. Many typical accidental fire scenarios have Q_d^* values on the order of 1. The physical diameter of a realistic fire may not be well-defined and may not actually matter when assessing the "size" of a fire. Instead, a characteristic diameter, D^* is considered in the definition of Q_d^* as noted in the table. The range of values of Q_d^* varied as a function of fuel type. In this study, Q_d^* took on values as small as 0.7 and as large as 44 as seen in Table 25.

The second entry in the table is the global equivalence ratio (ϕ), which is associated with the overall fire-induced ventilation and compartment stoichiometry. An estimate of the maximum achievable steady-state oxygen supply is given by: $\dot{m}_{O_2} = \dfrac{0.23}{2} A_o \sqrt{h_o}$, where $\dot{m}_{O_2}$ is an empirical correlation for the mass flow rate of oxygen (kg/s), A_o and h_o are the area and height of the doorway opening (m²), 0.23 is the mass fraction of oxygen in air. The parameter r in the table is the mass-based stoichiometric ratio of fuel to air required for complete combustion. The value of ϕ is useful in characterizing whether a given compartment fire is limited in size by its fuel supply or by its oxygen supply. The correlation for oxygen entrainment is valid for flashover conditions only, that is for values of $\phi > 1$.

In all of the experiments performed as part of this study, the fuel mass flow rate was either controlled (for the gaseous fuels) or measured (for the condensed fuels), whereas the oxygen supply was naturally controlled by the size of the compartment doorway and the fire heat release rate. The range of values of ϕ varied as a function of fuel type, taking on values as small as 0.01 as seen in Table 25. The value of ϕ was less than 1.0 for almost all of the experimental conditions, except for natural gas when ϕ was as large as 1.7. This implies that conditions inside of the compartment were nearly always over-ventilated. There is strong evidence that the estimate for air mass flow is over-predicted using this approach. For each of the fuels listed in the table, flames were observed outside of the doorway (see photos in section 3 of this report), oxygen volume fractions were near zero and increased CO production was measured in the upper layer for the largest fires sizes. These are all strong indicators of underventilated burning. This may be due to an inaccurate assumption of the incoming air mass flow rate, or an invalid assumption in the GER model that all the incoming air enters the mixed upper layer. Pitts [51] proposed that a large fraction of the incoming air is entrained into the out-flowing gases and never reaches the reaction zone.

The third entry in the table is the compartment height, H, normalized by D^*. The parameter H/D^* relates $\dot{Q}$ to its physical dimensions and indicates the relative importance of the fire plume to other features of the fire-driven flow, such as the ceiling jet or doorway flow. The range of values of H/D^* varied as a function of fuel type, taking on values as small as 1.4 and as large as 5.2 for conditions examined in this study, as seen in Table 25.

7 Conclusions

This reports details the test methods and experimental results from a series of fire tests in the reduced scale enclosure. The following list describes the main findings of this work:

- New measurements of total hydrocarbons, and soot were successfully performed. These new measurement provide a more complete data set for validating and improving predictive fire models.

- The performance of various burner designs, wall materials and sample conditioning methods were evaluated in order to aid in the planning of future full-scale fire experiments.

- A detailed analysis of thermocouple temperature measurements provided an estimate of radiative error and response time in the thermal environment of a fully developed enclosure fire (see Appendix A).

- CFD modeling was successfully used to examine the effects of probe interactions (see Appendix B) and help design the experiments.

- The gas species composition measurements showed that methane was the most abundant hydrocarbon species in the upper layer for all of the fuels and fire conditions tested, and was higher in concentration than the parent fuel in all cases.

- No significant amount of hydrocarbons was measured in the upper layer of the compartment in the toluene or polystyrene fires.

- As much as 60 % of the carbon in a polystyrene fire was present in the form of soot, which was more than twice the amount from any of the other hydrocarbons tested.

- The examination of compartment fire species data in terms of mixture fraction provided a rapid assessment of the overall chemical structure of the combustion environment. The measured data plotted as a function of mixture fraction confirmed the linear state relationship model for complete combustion when the fire was well ventilated.

- The results show that it is useful to consider soot as part of a mixture fraction analysis of compartment fires. The compartment gas species measurements showed that the CO concentration was a complex function of fuel type, fire size, ventilation and compartment sample location. The measured data plotted as a function of mixture fraction was consistent with previous studies that show the linear state relationship model is not valid for predicting CO in underventilated fires. In addition, a systematic correlation with temperature was not found. The results suggest that more research is needed to unravel the complexities of compartment fire chemistry.

- The species measurements showed that the soot yield was a function of fuel type and local equivalence ratio in the upper layer, whereas the ratio of the CO yield to the soot yield was independent of local equivalence ratio, and was not dissimilar to the results reported previously in Ref. [46].

- The mass fraction of CO in the upper layer of a reduced-scale compartment fire does not systematically correlate with the local temperature or mixture fraction, nor is it in local thermodynamic equilibrium.

- The yields of CO and soot in the exhaust stack were usually, but not always, lower than the local yields in the upper layer of a compartment, for a wide range of fuels and heat release rates.

- The post-compartment conditions generally had larger combustion efficiency values than local conditions in the upper layer of the reduced-scale compartment fires, for a wide range of fuels and heat release rates.

- The ratio of the yields of CO to soot in the upper layer of the reduced-scale compartment were relatively constant for each of the fuels types, for the fires burning heptane, toluene and polystyrene; in addition, the values were comparable to previous measurements [46] in the plumes of over-ventilated turbulent hydrocarbon fires.

- The yields of CO and soot in the upper layer of the reduced-scale compartment fires were related to the local equivalence ratio for fires burning ethanol, heptane, toluene and polystyrene, over a wide range of fire sizes, but significant scatter precludes use of the data for CO prediction; the relationship was non-linear in fires burning methanol and natural gas.

8 Future Work

The experiments described in this report are part of an ongoing project to explore the thermal and chemical phenomena associated with, primarily, underventilated compartment fires. In the next phase of this project fire measurements will be conducted using a full-scale ISO 9705 compartment with quantified uncertainties. Based on the second scaling rule shown in Table 25 and the results shown in Fig. 42, oxygen depletion could be expected in the front of an ISO 9705 enclosure for fires larger than about 1800 kW. Since investigation of underventilated burning is important, it may be advantageous to consider using a narrow doorway to reduce the fire size requirements to obtain conditions of interest in the experimental enclosure. The following tasks are planned to further this goal:

- Investigate the effects of distributed fuel sources on the structure of full-scale enclosure fires.

- Explore the effects of larger global equivalence ratios (primarily through decreased ventilation).

- Investigate scaling effects on the chemical and thermal structure of the fires.

- Explore the utility and durability of high-temperature-resistant blankets for full-scale enclosure construction.

- Improve the high temperature accuracy and time response of aspirated thermocouple probes through improved design.

- Extend gas species measurements to include quantification of hydrogen and water.

- Instrument the doorway to differentiate the heat release within and external to the compartment and to enable determination of the flows of enthalpy, mass and elemental carbon in and out of the compartment.

9 References

[1]   ISO 9705, Fire Tests - Full-Scale Room Test for Surface Products First Edition, International Organization for Standardization: Geneva, Switzerland (1993).

[2]   K.B.McGrattan, Fire Dynamics Simulator (Version 4): Technical Reference Guide., NIST SP 1018 (2004).

[3]   National Fire Protection Association and Society of Fire Protection Engineers, SFPE handbook of fire protection engineering, 3rd ed (2002).

[4]   C.L.Beyler, Major Species Production by Diffusion Flames in A 2-Layer Compartment Fire Environment, Fire Safety Journal 10 (1), 47-56 (1986).

[5]   E.E.Zukoski, J.H.Morehart, T.Kubota, and S.J.Toner, Species Production and Heat Release Rates in 2-Layered Natural-Gas Fires, Combustion and Flame 83 (3-4), 325-332 (1991).

[6]   M.Cleary and Kent JH, Modeling of species in hood fires by conditional moment closure, Combustion and Flame 143 (4)357-368 (2005).

[7]   S.Brohez, G.Marlair, and C.Delvosalle, Fire calorimetry relying on the use of the fire propagation apparatus. Part I: Early learning from use in Europe, Fire and Materials 30 (2), 131-149 (2006).

[8]   S.Brohez, G.Marlair, and C.Delvosalle, Fire calorimetry relying on the use of the fire propagation apparatus. Part II: Burning characteristics of selected chemical substances under fuel rich conditions, Fire and Materials 30 (1), 35-50 (2006).

[9]   N.P.Bryner, E.L.Johnson, and W.M.Pitts, Carbon Monoxide Production in Compartment Fires - Reduced-Scale Enclosure Test Facility, NIST IR 5568 (1994).

[10] W.M.Pitts, E.L.Johnsson, and N.P.Bryner, Carbon Monoxide Formation in Fires By High-Temperature Anaerobic Wood Pyrolysis, Twenty-Fifth Symposium (International) on Combustion1445-1462 (1994).

[11] B.Y.Lattimer and R.J.Roby, Carbon monoxide levels in structure fires: Effects of wood in the upper layer at a post-flashover compartment fire, Fire Technology 34 (4), 325-355 (1998).

[12] D.T.Gottuk, R.J.Roby, and C.L.Beyler, The role of temperature on carbon monoxide production in compartment fires, Fire Safety Journal 24 (4), 315-331 (1995).

[13] B.Y.Lattimer, U.Vandsburger, and R.J.Roby, Species transport from post-flashover fires, Fire Technology 41 (4), 235-254 (2005).

[14] R.G.Gann, J.D.Averill, E.L.Johnsson, M.R.Nyden, and R.D.Reacock, Smoke component yields from room-scale fire tests, NIST TN 1453 (2003).

[15] M.M.Hirschler, Analysis of work on smoke component yields from room-scale fire tests, Fire and Materials 29 (5), 303-314 (2005).

[16] P.Blomqvist and A.Lonnermark, Characterization of the combustion products in large-scale fire tests: Comparison of three experimental configurations, Fire and Materials 25 (2), 71-81 (2001).

[17] W.M.Pitts, The Global Equivalence Ratio Concept and the Formation Mechanisms of Carbon-Monoxide in Enclosure Fires, Progress in Energy and Combustion Science 21 (3), 197-237 (1995).

[18] G.Bertin, J.M.Most, and M.Coutin, Wall fire behavior in an under-ventilated room, Fire Safety Journal 37 (7), 615-630 (2002).

[19] A.Y.Snegirev, G.M.Makhviladze, V.A.Talalov, and A.V.Shamshin, Turbulent Diffusion Combustion under Conditions of Limited Ventilation: Flame Projection Through an Opening, Combustion, Explosion and Shock Waves 39 (1), 1-10 (2003).

[20] Y.Utiskul, J.G.Quintiere, A.S.Rangwala, B.A.Ringwelski, K.Wakatsuki, and T.Naruse, Compartment fire phenomena under limited ventilation, Fire Safety Journal 40 (4), 367-390 (2005).

[21] E.H.Yii, A.H.Buchanan, and C.M.Fleischmann, Simulating the effects of fuel type and geometry on post-flashover fire temperatures, Fire Safety Journal 41 (1), 62-75 (2006).

[22] E.H.Yii, C.M.Fleischmann, and A.H.Buchanan, Experimental study of fire compartment with door opening and roof opening, Fire and Materials 29 (5), 315-334 (2005).

[23] NIST, Final report on the collapse of the World Trade Center towers, NIST NCSTAR (2005).

[24] W.L.Grosshandler, N.P.Bryner, D.Madrzykowski, and K.Kuntz, Report of the technical investigation of The Station nightclub fire, NIST NCSTAR 2 (2005).

[25] D.Madrzykowski and W.D.Walton, Cook County Administration Building fire, 69 West Washington, Chicago, Illinois, October 17, NIST SP1021 (2004).

[26] A.Hamins, A.Maranghides, E.L.Johnson, M.K.Donnelly, J.C.Yang, G.W.Mulholland, and R.Anleitner, Report of Experimental Results for the International Fire Model Benchmarking and Validation Exercise #3, NIST SP1013-1 (2005).

[27] C.Huggett, Estimation of Rate of Heat Release by Means of Oxygen-Consumption Measurements, Fire and Materials 4 (2), 61-65 (1980).

[28] W.J.Parker, Calculations of the Heat Release Rate by Oxygen-Consumption for Various Applications, Journal of Fire Sciences 2 (5), 380-395 (1984).

[29] R.A.Bryant, T.J.Ohlemiller, E.L.Johnsson, A.Hamins, B.S.Grove, A.Maranghides, G.W.Mulholland, and W.F.Guthrie, The NIST 3 Megawatt Quantitative Heat Release Rate Facility - Description and Proceedure, NIST IR 7052 (2004).

[30] ASTM D 1945-03, Standard Test Method for Analysis of Natural Gas by Gas Chromatography, (2003).

[31] ASTM D 3588 - 98, Standard Practice for Calculating Heat Value, Compressibility Factor, and Relative Density of Gaseous Fuels, (2003).

[32] ASTM E 603-06e1, Standard Guide for Room Fire Experiments, (2006).

[33] L.G.Blevins and W.M.Pitts, Modeling of bare and aspirated thermocouples in compartment fires, Fire Safety Journal 33 (4), 239-259 (1999).

[34] G.E.Glawe, F.S.Simmons, and T.M.Stickney, Radiation and Recovery Corrections and Time Constants of Several Chromel-Alumel Thermocouple Probe in High Temperature, High Velocity Gas Streams, NACA TN3766 (1953).

[35] The Temperature Handbook, Omega Engineering Inc. 5th Edition (2004).

[36] B.J.McCaffrey and G.Heskestad, Robust Bidirectional Low-Velocity Probe for Flame and Fire Applications, Combustion and Flame 1125-127 (1976).

[37] R.Bryant, C.Womeldorf, E.Johnsson, and T.Ohlemiller, Radiative heat flux measurement uncertainty, Fire and Materials 27 (5), 209-222 (2003).

[38] W.M.Pitts, A.V.Murthy, J.L.de Ris, J.R.Filtz, K.Nygard, D.Smith, and I.Wetterlund, Round robin study of total heat flux gauge calibration at fire laboratories, Fire Safety Journal 41 (6), 459-475 (2006).

[39] R.W.Bilger, Reaction-Rates in Diffusion Flames, Combustion and Flame 30 (3), 277-284 (1977).

[40] N.Peters, Laminar Diffusion Flamelet Models in Non-Premixed Turbulent Combustion, Progress in Energy and Combustion Science 10 (3), 319-339 (1984).

[41] Y.R.Sivathanu and G.M.Faeth, Generalized State Relationships for Scalar Properties in Nonpremixed Hydrocarbon Air Flames, Combustion and Flame 82 (2), 211-230 (1990).

[42] A.Hamins and K.Seshadri, The Structure of Diffusion Flames Burning Pure, Binary, and Ternary Solutions of Methanol, Heptane, and Toluene, Combustion and Flame 68 (3), 295-307 (1987).

[43] J.E.Floyd, C.J.Wieczorek, and U.Vandsburger, Simulation of the Virginia Tech Fire Research Laboratory Using Large Eddy-Simulation With Mixture Fraction Chemistry and Finite Volume Radiative Heat Transfer, Proceedings of the 9th International Interflam Conference, Volume 1. September 17-19, 2001, Edinburgh, Scotland767-778 (2001).

124

[44] Y.R.Sivathanu and G.M.Faeth, Temperature Soot Volume Fraction Correlations in the Fuel-Rich Region of Buoyant Turbulent-Diffusion Flames, Combustion and Flame 81 (2), 150-165 (1990).

[45] W.C.Reynolds, The Element Potential Method for Chemical Equilibrium Analysis: Implementation in the Interactive Program STANJAN, Department of Mechanical Engineering Stanford University available at: http://navier.engr.colostate.edu/tools/equil.html. (1986).

[46] U.O.Koylu and G.M.Faeth, Carbon-Monoxide and Soot Emissions from Liquid-Fueled Buoyant Turbulent-Diffusion Flames, Combustion and Flame 87 (1), 61-76 (1991).

[47] R.Puri and R.J.Santoro, Proc. 3rd Int. Sym. Fire Safety Science595-604 (1991).

[48] A.Tewarson, F.H.Jiang, and T.Morikawa, Ventilation-Controlled Combustion of Polymers, Combustion and Flame 95 (1-2), 151-169 (1993).

[49] NIST Standard Reference Database Number 69, NIST Chemistry WebBook available at: http://webbook.nist.gov/chemistry/. (2005).

[50] J.G.Quintiere, Scaling Applications in Fire Research, Fire Safety Journal 15 (1), 3-29 (1989).

[51] W.M.Pitts, Toxic Yield, Technical Basis for Performance Based Fire Regulations., A Discussion of Capabilities, Needs and Benefits of Fire Safety Engineering. United Engineering Foundation Conference. Proceedings.76-87 (2001).

10 Acknowledgements

A number of individuals were instrumental in the completion of this work. The fabrication work and execution of these tests would not have been possible without the expert assistance and enthusiasm of the LFL staff: Laurean DeLauter, Edward Hnetkovsky and Jack Lee. Mike Selepak did exemplary work in carefully performing the labor intensive extractive soot measurements. Jay McElroy performed the video documentation and instrument installation. The authors are grateful to the following NIST staff for extremely helpful discussions that were a significant contribution to this study: Kevin McGratten, Jason Floyd, Rodney Bryant, Bill Pitts, Alex Maranghides, Bill Grosshandler, and Andrew Lock.

This work is dedicated in memory of Jack Lee (1953-2007). Jack was a technician in the large fire laboratory and will be greatly missed. He provided 35 years of loyal service to NIST and countless contributions to fire safety research.

11 Appendices

A. Analysis of Thermocouple Temperature Measurement

To estimate components of measurement uncertainty and instrument time response, the present study performed a series of detailed flow and heat transfer calculations, focusing on double shielded aspirated thermocouples and bare bead thermocouples. The character of the calculations is summarized in Table A1. There were two categories of calculations. The first category involved three-dimensional (3D) computational fluid dynamic (CFD) simulations, while the second involved an algebraic solution of the simplified energy balance equation. Two types of CFD modeling were performed. The first considered the realistic geometry in order to understand details of the flow field associated with the aspirated thermocouple. The second considered a simplified geometry, focusing on details of the heat transfer process. Ideally, the CFD simulation would be able to consider details of the flow and heat transfer in the actual geometry, including turbulence and conjugate heat transfer with conduction, convection and radiation for the real geometry, but this would be prohibitively expensive computationally. For this reason, the CFD calculations were split into detailed flow calculations for the actual geometry, and detailed conjugate heat transfer calculations for a simplified geometry. The 3D CFD calculation results are compared with algebraic solutions of a simplified energy balance model. The results predicted by the algebraic model were in agreement with the CFD model over a broad temperature range, despite its many assumptions and idealizations. A parametric study was conducted to quantify the thermocouple errors for various gas temperature and surrounding conditions. The CFD solutions were transient, allowing determination of instrument time response. This is in contrast to the analytical solution, which presented a steady-state solution only.

Table A1. Summary of the Numerical Simulations.

	Simple energy balance model	3D CFD model for Simple Geometry	3D CFD model for Real Geometry
Physics	Heat transfer based on specified aspiration rate	Flow and heat transfer	Turbulence associated with aspiration flow
Solid Phase Heat Transfer	Thermally Thin	Considered	Not Considered
Radiative Heat Transfer	Surface radiation	Surface-Surface Radiation	Not Considered
Convective Heat Transfer	Empirical correlation	Surface Energy balance	Surface Energy Balance
Surface Emissivity	0.8	0.8	Not considered
Surrounding Temp.	300 K to 1200 K	300 K to 1200 K	Adiabatic
Gas Temp.	300 K to 1200 K	300 K to 1200 K	300 K, 1200 K
Solution Type	Steady Solution	Transient solution	Transient Solution
Turbulence	Empirical correlation	Standard K-ε model	Standard K-ε model
External Gas Velocity	1 m/s	1 m/s	1 m/s
No. Grids	None	25000	1500000

3D Flow Modeling for a Realistic Geometry

The 3D flow modeling considered the realistic geometry of an open-ended double shielded aspirated thermocouple as seen in Fig. A1. The computational configuration included an inner and outer shields, a bare type K thermocouple taken to be cylindrical, and an extended domain. The computational domain was divided into approximately 1.5 million cells of a tetrahedral type mesh using the ICEM-CFD [2], which is a commercial CAD and grid generation program. Figures A2 and A3 show the computational grids for the open-end and open-side aspirated thermocouple geometries. Both types were used in the RSE compartment fire experiments. The flow fields were calculated using a commercially available CFD package to model the flow for given operating conditions [3]. The code is based on the finite volume method on a collocated grid. A non-staggered grid system was used for storage of discrete velocities and pressures. The standard k-ε turbulence model and incompressible ideal gas assumption were applied to solve the Reynolds stress term and the density change, respectively. The governing equations were discretized by the 2nd order upwind scheme in space and the SIMPLE algorithm (Semi-Implicit Method for Pressure Linked Equation) with under-relaxation used to iteratively solve the momentum equation in discretized form. The implicit solver was used to capture the main features of the unsteady motion around the aspirated thermocouples. In this study, CFD calculations investigated the flow field in the inner and outer aspiration tubes with an emphasis on characterizing any differences. That information was then used as boundary conditions for the CFD and analytic heat transfer analyses.

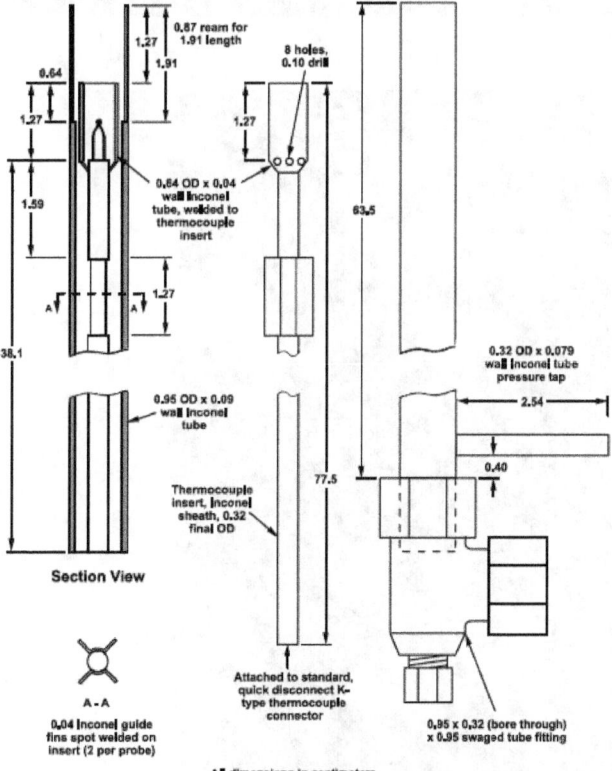

Figure A1. Schematic of a double shielded end-open aspirated thermocouple.

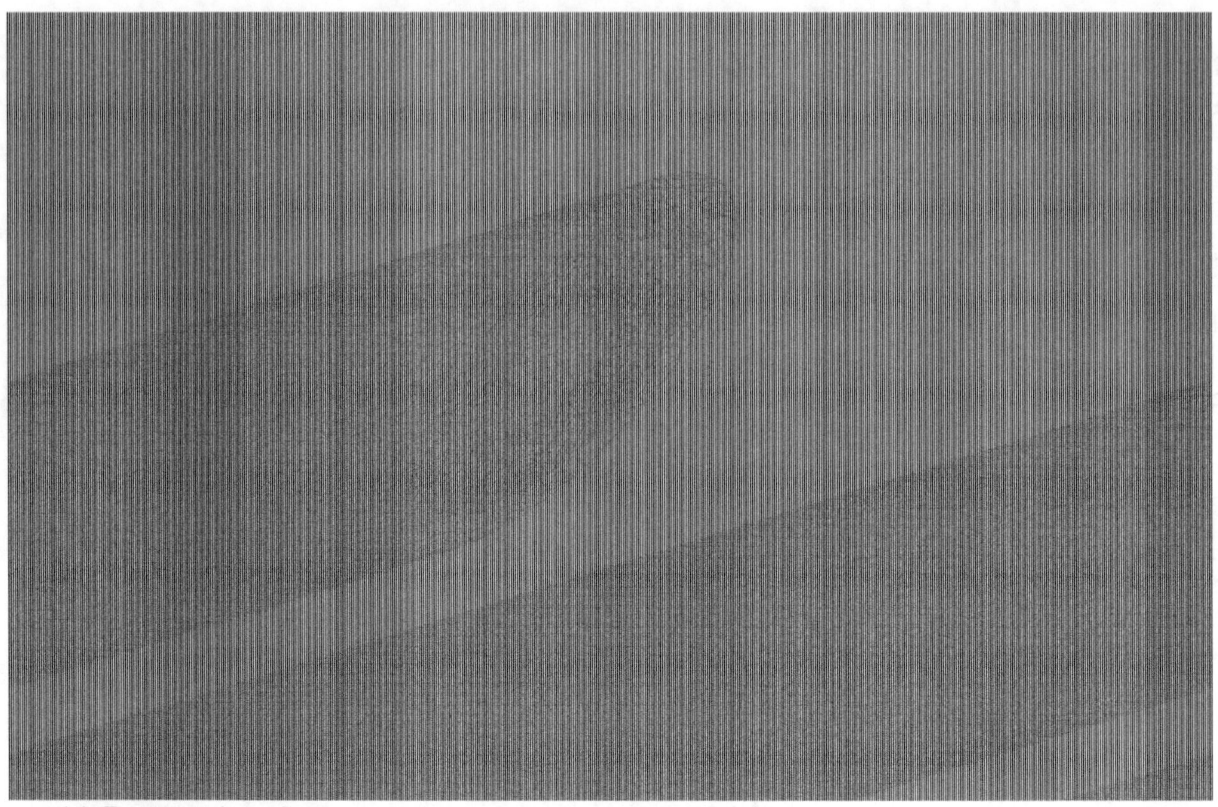

(a) Perspective view

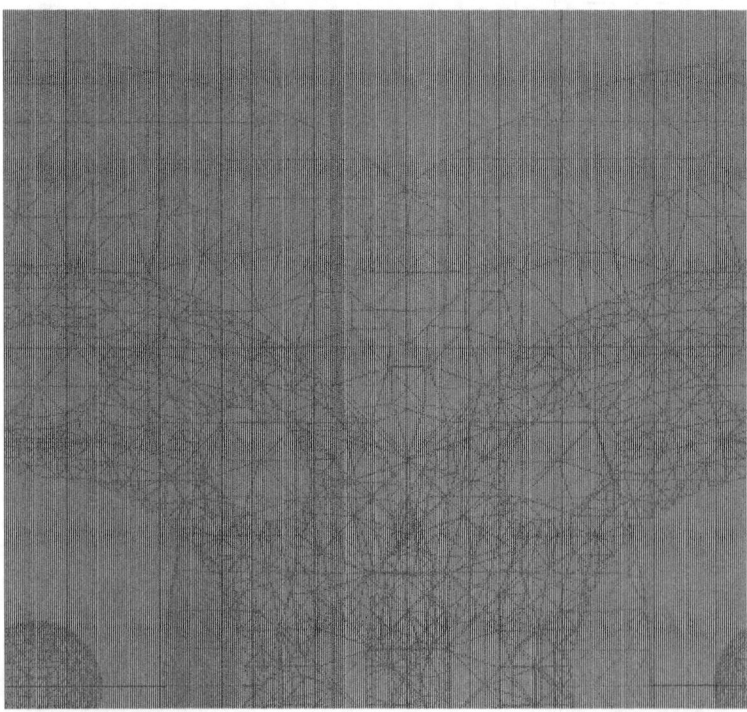

(b) Front view

Figure A2. Computational grids for end-open double shield aspirated thermocouples.

(a) Perspective view

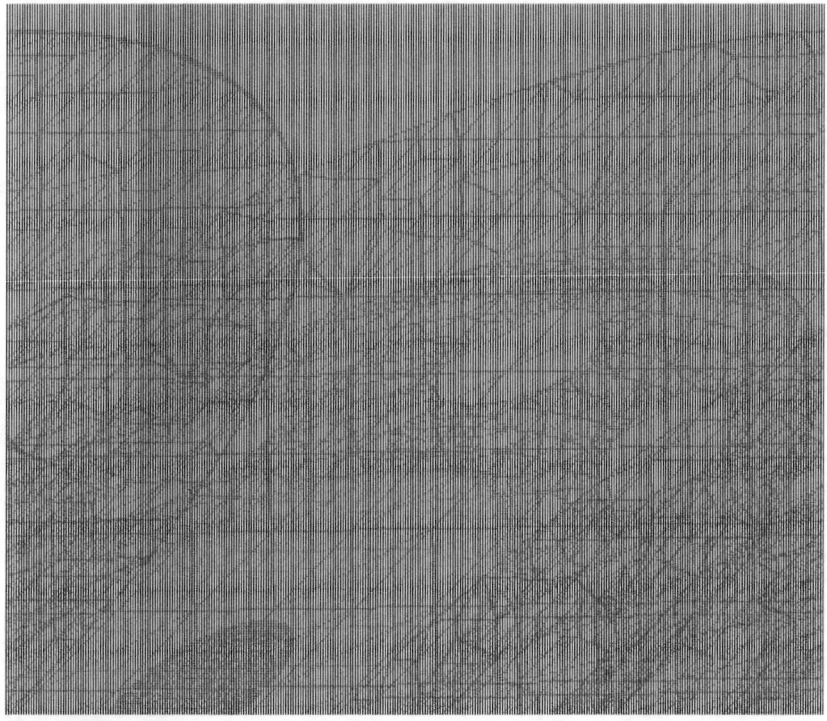

(b) Front view

Figure A3. Computational grids for side-open double shield aspirated thermocouples.

129

Conjugate Heat Transfer Modeling for a Simplified Geometry

A detailed 3D heat transfer calculation (including conduction, convection, and radiation) was performed to estimate the effectiveness of bare-bead and double-shielded aspirated thermocouples for a simplified geometry. Figure A4 presents a schematic of the two configurations. The double-shielded calculation assumed that the flow domain consisted of an external flow, an annular flow between the outer and inner cylinders, and an inner flow within the inner cylinder. In the bare-bead calculation, the thermocouples were directly exposed to the external gas and the surroundings without a shield or an applied aspiration flow. The incoming gas temperature was assumed to be uniform. The incoming flow velocity induced by aspiration was imposed as a boundary condition, and was determined from the detailed flow calculations described above. The material properties of the shield and thermocouple bead were taken to be steel and nickel, respectively, and are listed in Table A2. This is a reasonable approximation as K-type thermocouples are composed of more than 90 % nickel [4].

Table A2. Material Properties [5].

Material	Density (kg/m^3)	Specific heat (J/kg·K)	Thermal conductivity (W/m·K)
Nickel	8900	460	91.7
Steel	8030	502	16.3

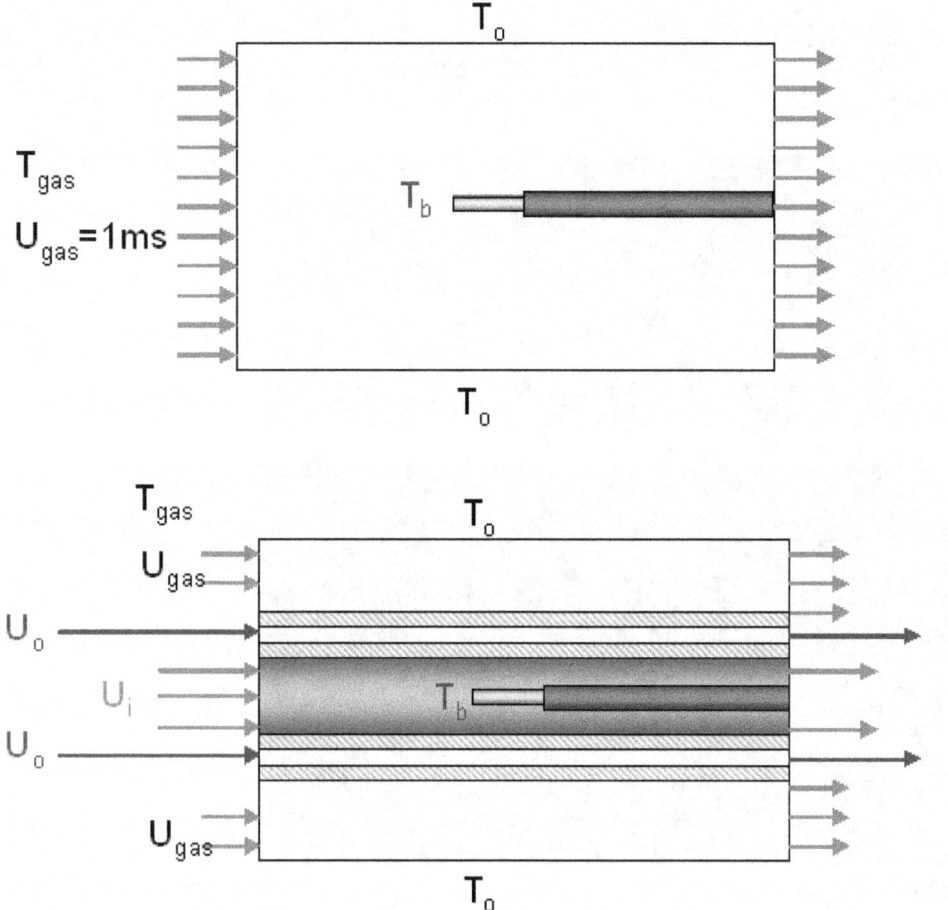

Figure A4. Schematic of bare bead and open-end double shielded aspirated thermocouples.

Radiative heat transfer was computed using a surface to surface radiation model in which the energy exchange between the two surfaces depended on the view factor, which is a geometric function involving the size, distance, and orientation of surfaces. The surfaces were taken as gray and diffuse. The calculations were performed assuming a constant emissivity (ε) of 0.8 for all metal surfaces, using the value recommended by Blevins [1]. The external flow velocity was assumed to be 1 m/s in all cases.

The representative thermocouple temperature (T_b) was calculated using a volume weighted average as follows:

$$T_b = \frac{1}{V_b} \int T \cdot dv = \frac{1}{V_b} \sum_{i=1}^{N} T_{b,i} \cdot dV_i \tag{A1}$$

The thermocouple effectiveness was defined in terms of the percentage error (E) between the incoming gas flow temperature (T_g in Fig. A4) and the representative temperature at the thermocouple bead (T_b in Fig. A4):

$$E = 100 \times \left(\frac{T_b - T_g}{T_g} \right) \tag{A2}$$

Steady-State Solution of the Algebraic Energy Balance Equation
An analytic solution of the steady state energy balance for aspirated thermocouples was previously reported [1]. That calculation, however, did not consider conductive heat transfer through the solid shield. Also, the temperature difference between the inner and outer surfaces was assumed to be zero. In the case of the double shielded aspirated thermocouple, the inner and outer gas velocities were assumed to be equal and the external flow was assumed to be parallel to the probe axis.

Results
Figure A5 shows the calculated pressure and velocity fields of an end-open double shielded aspirated thermocouple at ambient temperature with an aspiration flow rate of 24 L/min. The entrance area of the inner cylinder of the aspirated thermocouple was larger than the exit area through its eight inner holes (see Figs. A1 and A2), causing a relatively high stagnation pressure inside the inner shield. This adverse pressure gradient forced the aspiration flow to pass through the outer passage with a differentially high velocity as compared to the inner cylinder. The flow blockage effect was characterized as a ratio of the average velocities between the inner and outer annular passages. It influenced the rate of convective heat transfer in the aspirated thermocouple, impacting the effectiveness of the aspirated thermocouple. Figure A6 presents the calculated maximum values of the velocity of the inner shield (U_i), the outer annular passage (U_o), and the maximum value of the ratio of these velocities (ζ) as a function of the total aspiration flow. The velocity ratio (ζ) is defined as:

$$\varsigma = \frac{U_o}{U_i} \tag{A3}$$

For nominal operating conditions (24 L/min at STP), the maximum velocity in the inner shield was found to be about 3 times larger than that within the annular passage. Figure A6 shows that this ratio increased with aspiration flow.

pressure (pa)

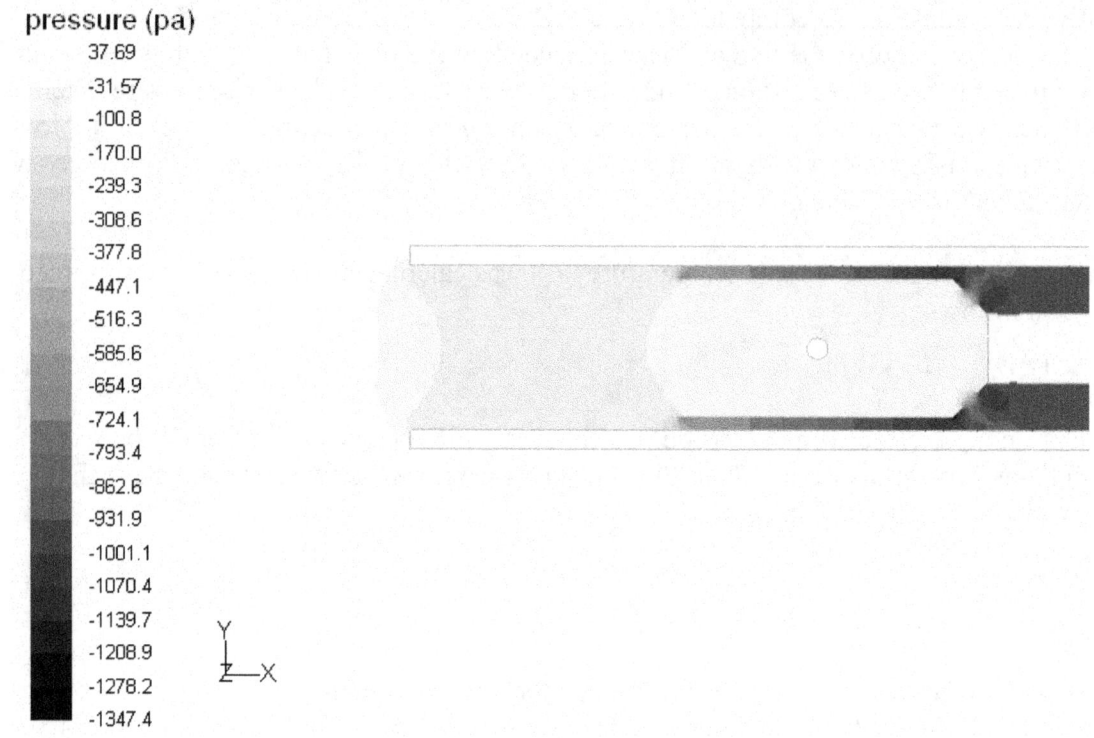

(a) Pressure field

velocity (m/s)

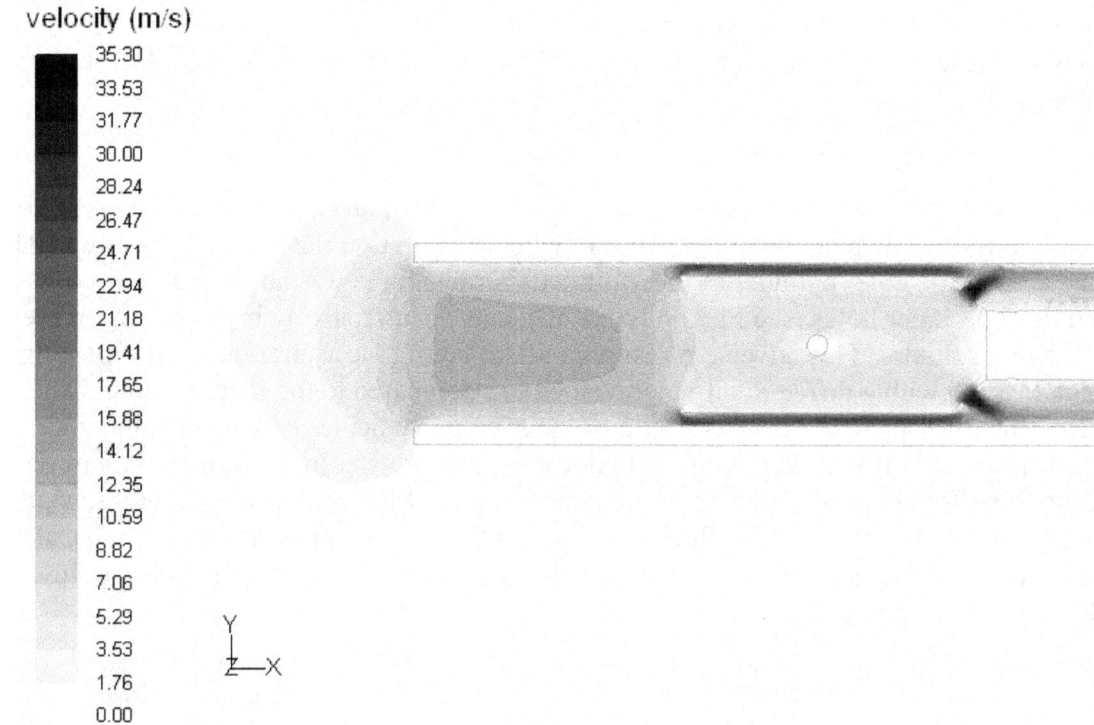

(b) Velocity magnitude

Figure A5. Calculated pressure and velocity fields in end-open double shielded aspirated thermocouples at ambient temperature with an aspiration flow rate of 24 L/min.

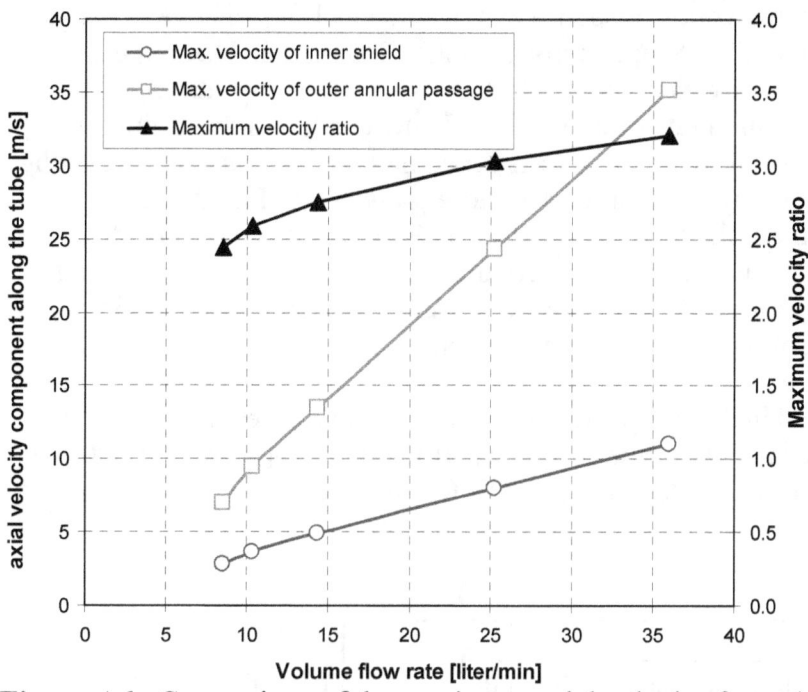

Figure A6. Comparison of the maximum axial velocity for end-open double shielded aspirated thermocouples (T_{gas} = 300 K).

Figure A7 shows the calculated thermocouple error (see Eq. A2) for a bare bead thermocouple as a function of the surrounding (T_o in Fig. A4) temperatures for both the simple energy balance model and the 3D CFD model for an incoming gas temperature of either 300 K or 900 K. The external gas velocity was taken as 1 m/s directed toward the open end of the aspirated thermocouple. The results show one large error regime of thermocouple measurement associated with low values of the gas temperature and high surrounding temperatures. The results show that the error was relatively small for moderately-high gas temperatures, regardless of the surrounding temperature. Despite its many assumptions, the solution to the simple energy balance model was essentially in agreement with the 3D CFD calculation results for the bare bead thermocouple.

Figure A8 show maps of the calculated thermocouple error (Eq. A2) for an open-end double-shielded aspirated thermocouple as a function of the thermocouple bead (T_b) and the surrounding (T_o) temperatures determined using the 3D CFD model. The map shows two regimes of significant error for the thermocouple temperature measurements. The first occurs for relatively low temperature surroundings in which the gas temperature is systematically under-predicted, notably for higher gas temperatures. The temperature error predicted by the CFD model in this regime varied with temperature, but was as large as -280 K for gas temperatures of 1200 K, corresponding to an error of about 25 %. The other significant-error regime occurs for large surrounding temperatures, in which the gas temperature is over-predicted. The maximum error in this regime was as large as 140 K for a high surrounding temperature (T_o = 1500 K) and a thermocouple temperature of about 900 K. Calculations using the algebraic model show that similar trends result, but the magnitude of the differences was much smaller. There is less confidence in the algebraic model, due to its many assumptions.

Figure A9 shows the calculated thermocouple response to a hypothetical step change in the gas temperature with an invariant surrounding temperature using the CFD model. The initial gas temperature and surrounding temperature were identical. Results are shown for initial temperatures of 1200 K and an aspiration rate of 24 L/min. The temperature rose or fell towards the final temperature over tens of seconds, the exact time depending on the relatively size of the step change. Results from several calculations similar to those presented in Fig. A9 are summarized in Fig. A10, which shows the time to reach a quasi-steady state condition after the hypothetical step change for various surrounding temperatures. Steady state was defined as the time when the thermocouple temperature variation was less than 1 K/s. Under this condition, the calculated results are within 10 K of the asymptotic calculation results.

The time to steady state varied with incoming gas temperatures, but was largest for high incoming gas temperatures. For example, for the case of a surrounding temperature of 300 K, the time to steady state took a maximum value of about 50 s for an incoming gas temperature of 1100 K.

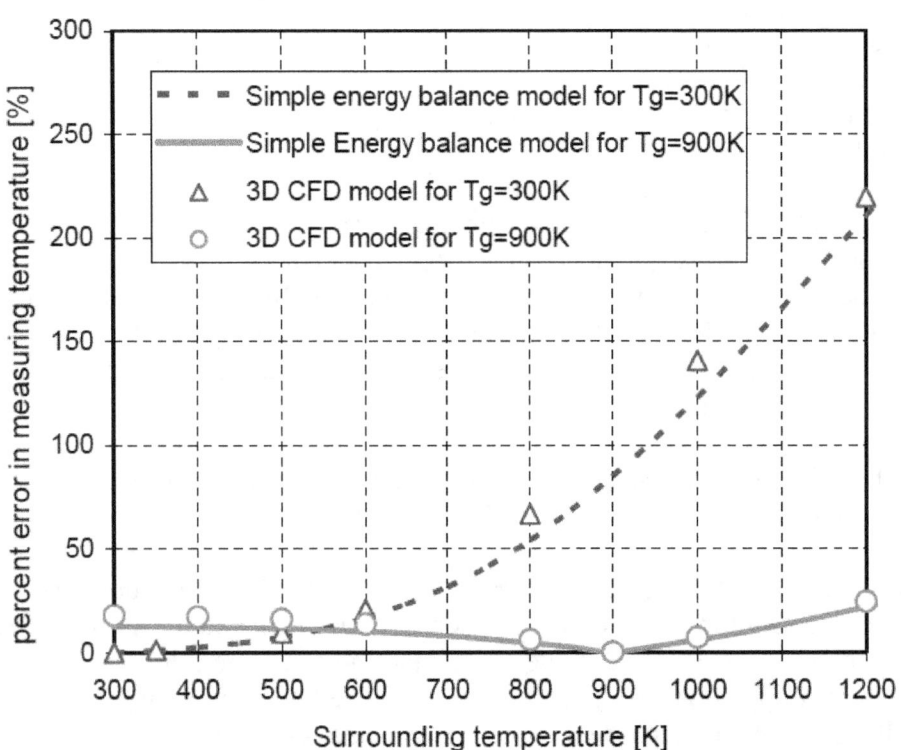

Figure A7. A comparison of the predicted percent error in measuring temperature between the simple energy balance model and the 3D CFD model for bare bead thermocouples with an external gas velocity of 1 m/s.

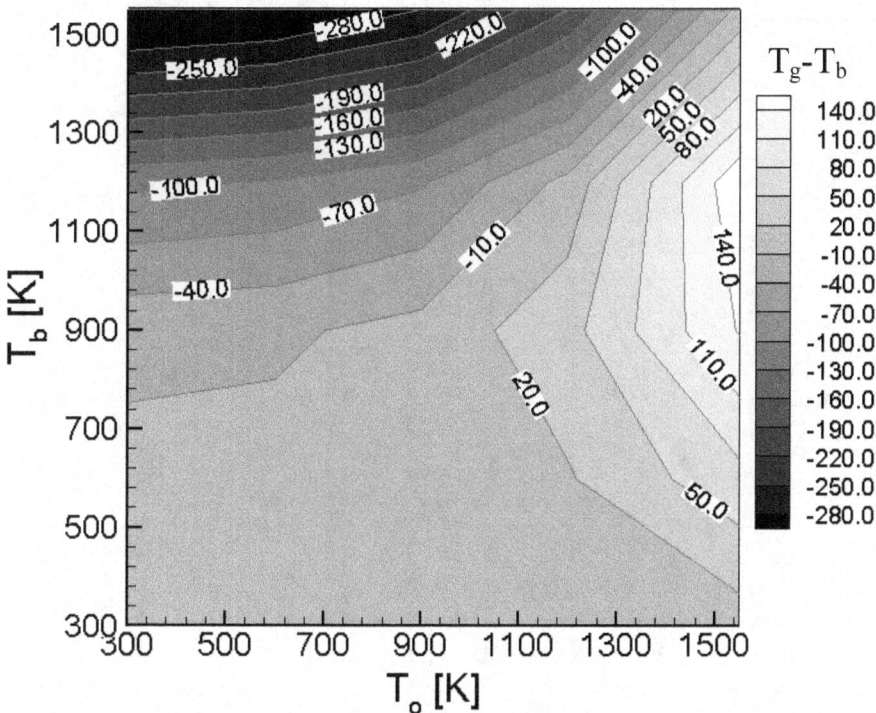

Figure A8. Map of the thermocouple uncertainty for a double shield aspirated thermocouple as a function of the measured (T_b) and surrounding (T_o) temperatures calculated using the CFD model.

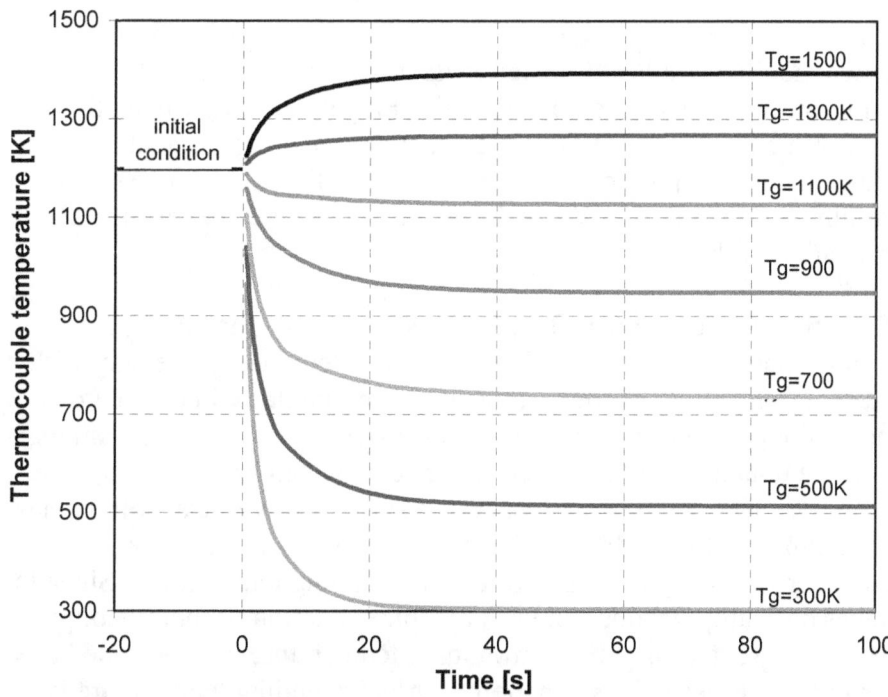

Figure A9. Calculated time history of the temperature of a double shielded aspirated thermocouple for various incoming gas temperatures (T_g) with a surrounding temperature of 1200 K and an aspiration flow of 24 L/min).

135

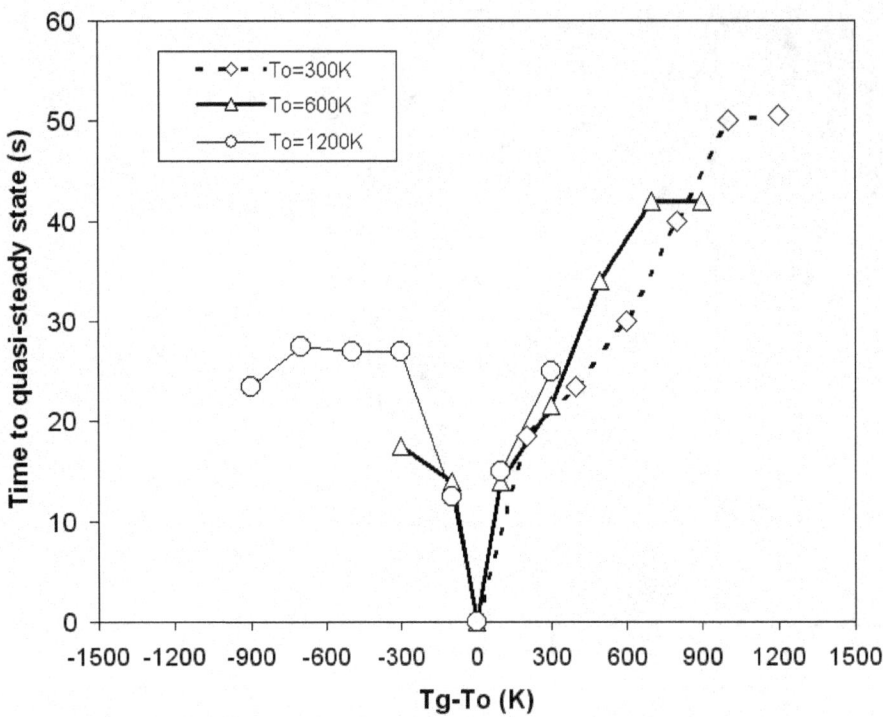

Figure A10. Comparison of calculated time to reach steady state for a double shielded aspirated thermocouple with a 24 L/min aspiration flow rate as a function of the incoming gas temperature for surrounding temperatures of 300 K, 600 K, and 1200 K.

Summary

The present study investigated the flow and heat transfer characteristics of aspirated thermocouples using a simple energy balance model and a 3D CFD model. The calculations quantify the systematic measurement error, providing an estimate of the measurement uncertainty. At the same time, the model provides information on the time to achieve steady-state, which should be carefully considered in terms of the development of an experimental procedure and in the interpretation of the results.

Despite the application of additional assumptions and idealizations, calculations using the previously developed algebraic energy balance model generally showed good agreement with the results of the 3D CFD model. The algebraic model can be useful, particularly in parametric studies used to evaluate thermocouple measurement error. Consistent with previous findings, calculations show that use of the double shield aspirated thermocouple can greatly reduce the thermocouple error especially for low gas temperatures. The results, however, can still be biased by hundreds of degrees, depending on the conditions. Figure A8 provides information on the magnitude of the measurement error for a given value of the surrounding temperature. Since the surrounding temperature takes on multiple values and is a complex function of thermocouple location and fire conditions, a representation of the surrounding temperature can be made based on estimated temperature averages. Precise determination of the surrounding temperature is impossible, and engineering judgment is a key component of the uncertainty analysis.

The results of the CFD model allow determination of the transient response of the double shield aspirated thermocouple, which is helpful in the interpretation of measurement results and possibly for design of the experiment itself. The calculated time response of the aspirated thermocouples suggests that while the measurements provide an adequate representation of the average local temperature, they do not provide an accurate representation of the magnitude of the temperature excursions. Turning the aspiration flow off during the experiment may bias the results if the time response is not taken into account. As a rule of thumb, for incoming gas temperatures less than 900 K, an aspirated thermocouple should be in place for at least ½ min before the data should be considered acceptable. For incoming gas temperatures greater than 900 K, an aspirated thermocouple should be in place for about 1 min before the data should be considered acceptable, unless the thermocouple is sampling from an optically thick upper layer, when ½ min should be adequate.

References
[1] Blevins, L.G., Behavior of bare and aspirated thermocouples in compartment fires, Proc. 33rd National Heat Transfer Conference, New Mexico, August 15-17, 1999.
[2] ICEM-CFD Version 4.0, User's Manual.
[3] FLUENT Inc., FLUENT Version 6.0 User's Guide.
[4] Burns, G. W. and Scroger, M. G., The calibration of thermocouples and thermocouple materials, NIST Special Publication 250-35, 1989.
[5] Incropera, F. P and DeWitt, D. P., Introduction to Heat Transfer, 2nd edition, Willy, 1993.

B. Analysis of Probe Interactions

There were five different types of measurement probes at various positions in the fire compartment. They included bi-directional probes for velocity, bare-bead and aspirated thermocouples for temperature, gas sampling tubes for gas species and soot sampling probes for the gravimetric soot measurement. These probes affect the flow field through geometric obstructions and create a flow sink due to the extraction of combustion gas. The effects of the probes on the flow field were not avoidable in the measurement system, but should be minimized during the test. Among the measurement probes, the suction type probes, such as the aspirated thermocouples, gas and soot sampling probes could significantly influence the flow field. If the suction type probes are operated too close to each other, there could be an interaction associated with the suction flow rate.

Before the tests were conducted, numerical simulations were used to investigate the effect of the sampling probe on the flow field for various operating conditions. The interactions between the aspirated thermocouples and the gas sampling probes were investigated because of the high suction flow rate. The experimental suction flow rate of the soot probe was similar to the gas sampling probe ($Q \approx 3$ L/min @ 300 K).

Figure B1 shows the configuration of the two sampling probes perpendicular to incoming gas flow. The flow field near the entrance of the sampling probe was examined for the typical operating conditions of the probes ($Q_1 = 30$ L/min for the aspirated thermocouple, $Q_2 = 3$ L/min for the gas sampling). In the calculations the incoming gas had a constant velocity (U_g) and gas temperature (T_g). Figure B2 shows the velocity field near the sampling probe for an incoming gas velocity of 1 m/s, and a gas temperature of 900 K. The influence of the sampling flow rate was restricted to the probe entrance. Generally, the gas velocities induced by the fire were much higher than the suction velocities of the sampling probes, and the effect of the probe suction on the flow field was negligible.

Figure B3 shows the streamlines at the center-plane of the probe. The distance from the probe entrance to the aspirated flow field disturbance was on the order of one diameter. Figure B4 and Fig. B5 represent the static pressure and stream-wise velocity (U_x) profiles along the center axis of the probes for the 1 m/s gas velocity and 900 K temperature. The static pressure and stream-wise velocity do not vary significantly in this configuration. Also, the velocity profiles of the aspirated thermocouple and the gas sampling probe have nearly the same magnitude. If one probe affects the flow field of the other probe, the velocity profiles would not match. The incoming gas flow dominates the entire flow field except at the entrance of the probes because the flow rate of the sampling probes is relatively small compared to the main gas flow. Therefore, the probe interactions were negligible in the main plume of the fire induced flow, but in a low velocity region such as stagnation or recirculation zone, the suction from one probe can affect the flow field and the measurement volume being interrogated by other probes.

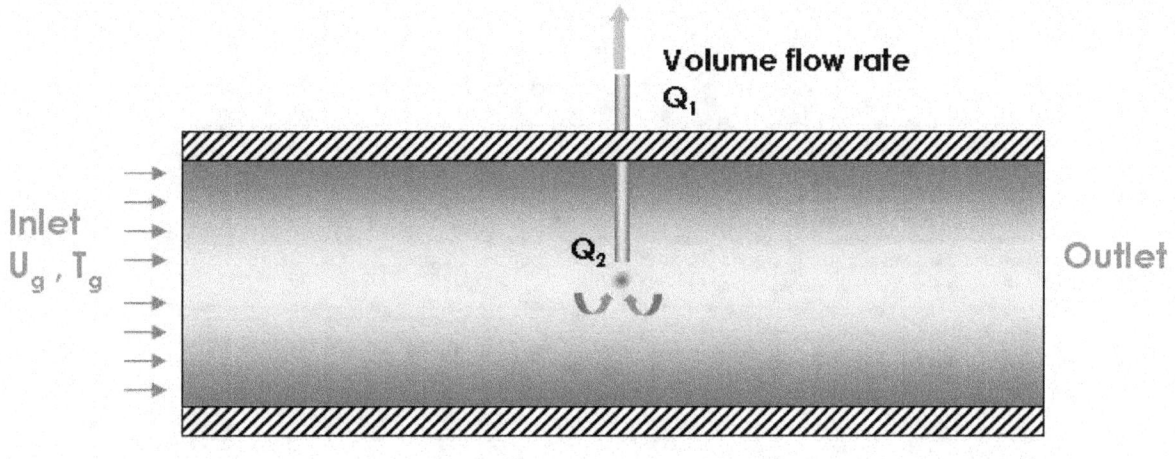

Figure B1. Schematic of the calculations of the sampling probes interaction.

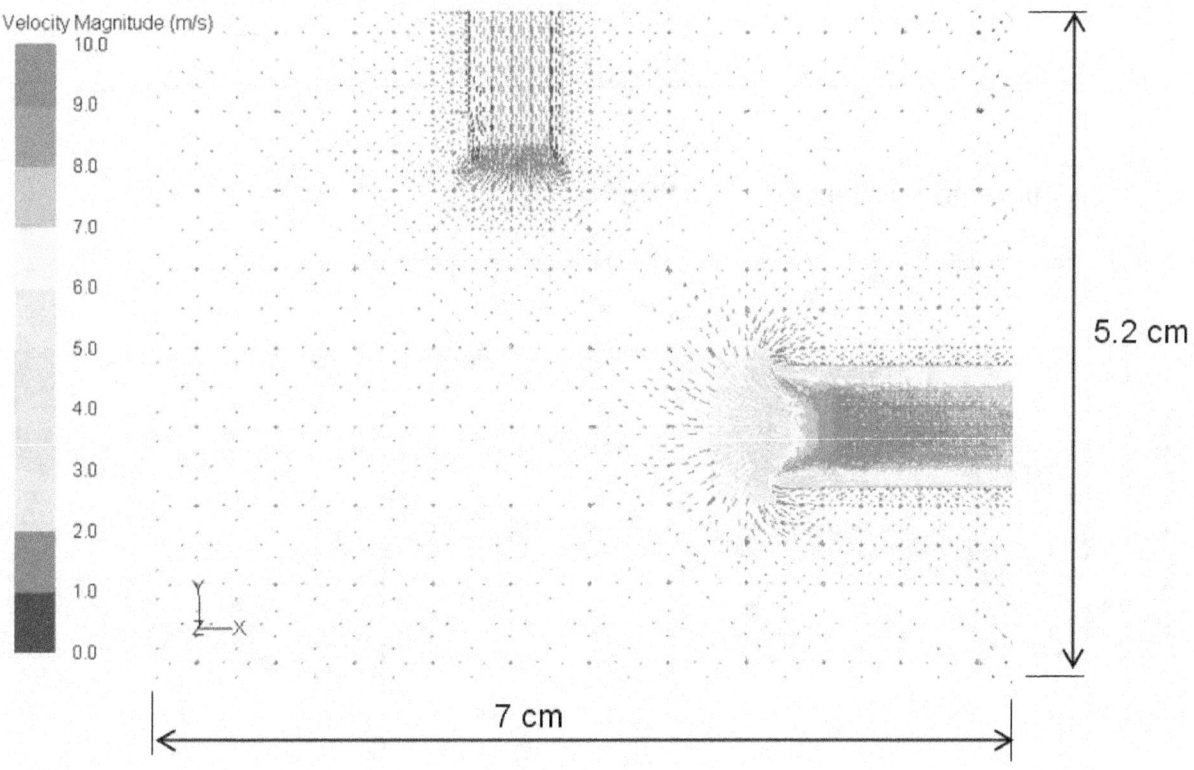

Figure B2. Velocity field across cross-section of the probes position.

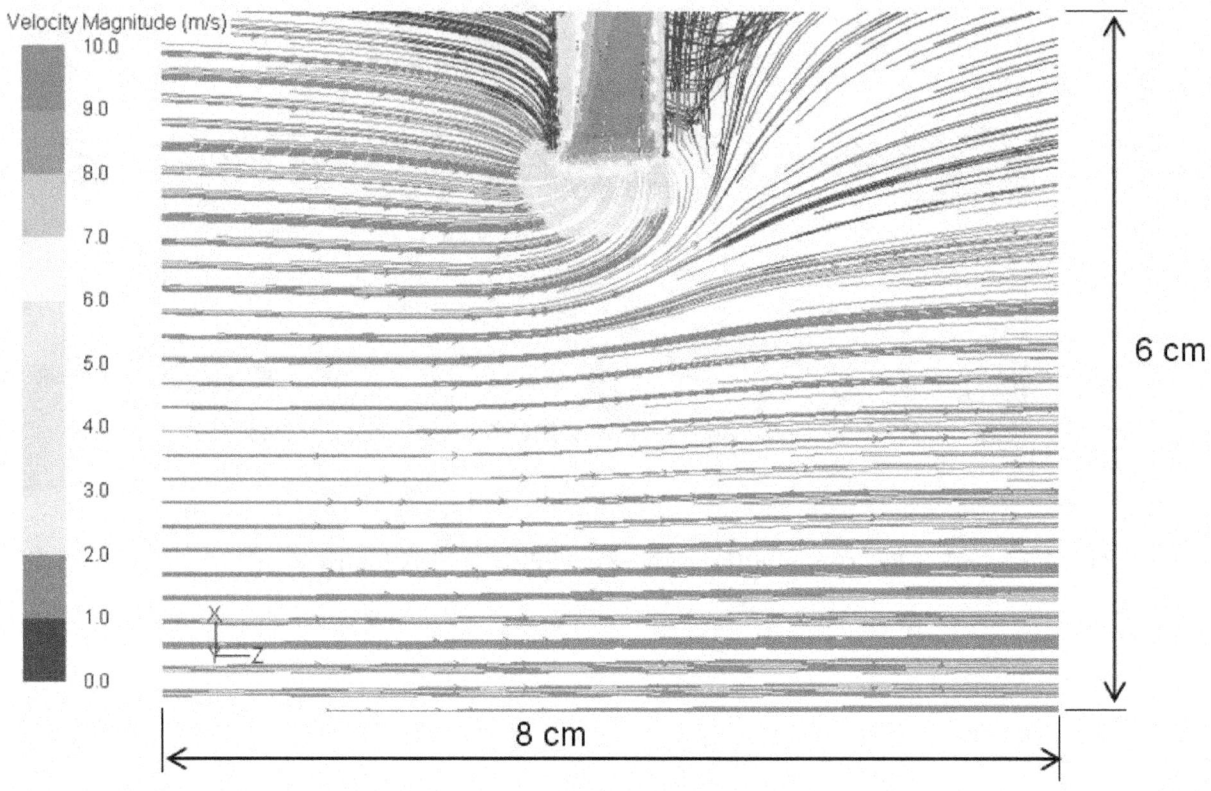

Figure B3. Streamlines near the aspirated thermocouple.

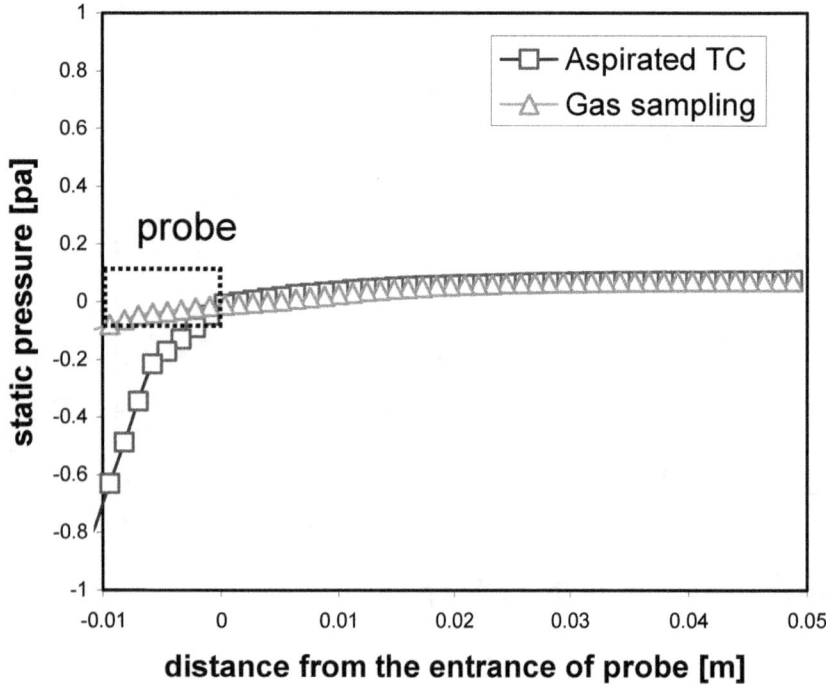

Figure B4. Static pressure profiles along the center axis of the probes.

inside of probe   outside of probe

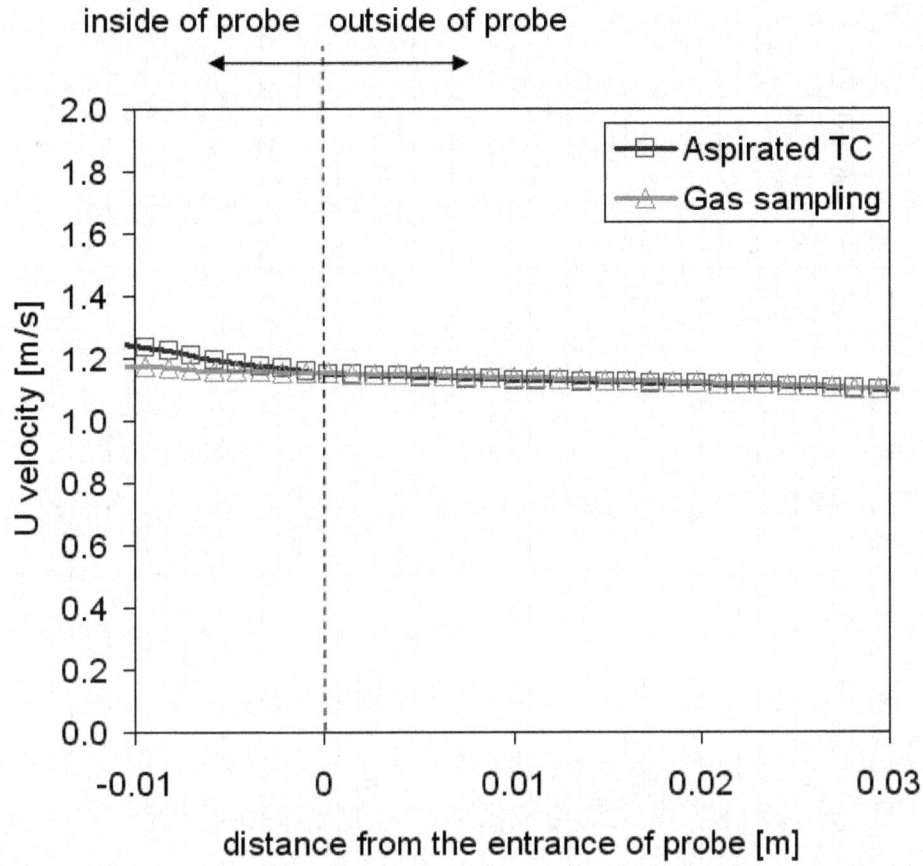

Figure B5. U-velocity profile along the center axis of the probes.

C. Channel Lists

LFL MIDAS Hookup Sheet Instrument and Channel Description — Series: — Underventilated RSE — Revision Date: 3/29/2006

Measurements	x x0=inside left wall (looking at door)	y y0=inside front (door) wall	z z0=floor	Location Description	Overall Channel Number	Abbr.	MIDAS Station	Module: Channels	Mod.	Mod. Ch. No.	Conv.Units	Wire	Gain
Main Channels													
	RSEmax=(95,142,98)												
O2 at Rear Sampling location	29	113	88	10 cm from C, 29 cm from Left, 29 cm from Rear	0	O2Rear	East	1:0-31	1	0	Vol Fr	Cu	1
Bad channel			88		1	Bad			1	1			1
CO2 at Rear Sampling location	29	113	88	10 cm from C, 29 cm from Left, 29 cm from Rear	2	CO2Rear	East		1	2	Vol Fr	Cu	1
CO at Rear Sampling location	29	113	88	10 cm from C, 29 cm from Left, 29 cm from Rear	3	CORear	East		1	3	Vol Fr	Cu	1
UH at Rear Sampling location	29	113	88	10 cm from C, 29 cm from Left, 29 cm from Rear	4	UHRear	East		1	4	Vol Fr	Cu	1
Rear Gas Rack Dewpoint				outside RSE	5	DPRear	East		1	5	°C	Cu	1
Rear Sampling location gas temperature 1-near probe				outside RSE	6	TGasRear1	East		1	6	°C	TC	100
Rear Sampling location gas temperature 2-near UH				outside RSE	7	TGasRear2	East		1	7	°C	TC	100
Rear Sampling location water inlet temperature				outside RSE	8	TH2OInRear	East		1	8	°C	TC	100
Rear Sampling location water outlet temperature				outside RSE	9	TH2OOutRear	East		1	9	°C	TC	100
O2 at Front Sampling location	29	10	88	10 cm from C, 29 cm from Left, 10 cm from Front	10	O2Front	East		1	10	Vol Fr	Cu	1
CO2 at Front Sampling location	29	10	88	10 cm from C, 29 cm from Left, 10 cm from Front	11	CO2Front	East		1	11	Vol Fr	Cu	1
CO at Front Sampling location	29	10	88	10 cm from C, 29 cm from Left, 10 cm from Front	12	COFront	East		1	12	Vol Fr	Cu	1
UH at Front Sampling location	29	10	88	10 cm from C, 29 cm from Left, 10 cm from Front	13	UHFront	East		1	13	Vol Fr	Cu	1
Front Gas Rack Dewpoint				outside RSE	14	DPFront	East		1	14	°C	Cu	1
Front Sampling location gas temperature 1- near probe				outside RSE	15	TGasFront1	East		1	15	°C	TC	100
Front Sampling location gas temperature 2- near UH				outside RSE	16	TGasFront2	East		1	16	°C	TC	100
Front Sampling location water inlet temperature				outside RSE	17	TH2OInFront	East		1	17	°C	TC	100
Front Sampling location water outlet temperature				outside RSE	18	TH2OOutFront	East		1	18	°C	TC	100
Optical Soot (U/to in adj file)	48	-5	75	6 cm from Soffit, doorway CL	28	OptSoot	East		1	28	V	Cu	1
Optical Soot Laser Reference Signal	48	-5	75	6 cm from Soffit, doorway CL	29	OptSootRef	East		1	29	V	Cu	1
Optical Soot Apparatus Temperature				outside RSE	30	TOptSoot	East		1	30	°C	TC	100
Gravimetric Soot Probe 1 Filter Temperature - rear				outside RSE	31	TGravSoot1	East	2:0-31	1	31	°C	TC	100
Gravimetric Soot Probe 2 Filter Temperature -rear				outside RSE	32	TGravSoot2	East		2	0	°C	TC	100
Gravimetric Soot Probe 1 Solenoid Signal -rear				outside RSE	33	VGravSoot1	East		2	1	V	Cu	1
Gravimetric Soot Probe 2 Solenoid Signal -rear				outside RSE	34	VGravSoot2	East		2	2	V	Cu	1
Gravimetric Soot Probe 1 Mass Flow -rear				outside RSE	35	SootMF1	East		2	3	V	Cu	1
Gravimetric Soot Probe 2 Mass Flow-front				outside RSE	36	SootMF2	East		2	4	V	Cu	1
Total Heat Flux Gauge Front (SN=131836)	48	35	0	In floor on room CL, 35 cm from Front	37	HFF	East		2	5	kW/m2	Cu	100
Total Heat Flux Gauge Rear (SN=131835)	48	106	0	In floor on room CL, 35 cm from Rear	38	HFR	East		2	6	kW/m2	Cu	100
Doorway Pressure 79 cm Left (+out)	32	-5	79	79 cm from floor, doorway 8 cm from Left side of dr	39	PD79L	East		2	7	Pa	Cu	1
Doorway Pressure 79 cm CL (+out)	48	-5	79	79 cm from floor, doorway CL	40	PD79C	East		2	8	Pa	Cu	1
Doorway Pressure 79 cm Right (+out)	64	-5	79	79 cm from floor, doorway 8 cm from Right side of dr	41	PD79R	East		2	9	Pa	Cu	1
Doorway Pressure 60 cm (+out)	48	-5	60	60 cm from floor, doorway CL	42	PD60C	East		2	10	Pa	Cu	1
Doorway Pressure 40 cm (+in)	48	-5	40	40 cm from floor, doorway CL	43	PD40C	East		2	11	Pa	Cu	1
Doorway Pressure 20 cm Left (+in)	32	-5	20	20 cm from floor, doorway 8 cm from Left side of dr	44	PD20L	East		2	12	Pa	Cu	1
Doorway Pressure 20 cm CL (+in)	48	-5	20	20 cm from floor, doorway CL	45	PD20C	East		2	13	Pa	Cu	1
Doorway Pressure 20 cm Right (+in)	64	-5	20	20 cm from floor, doorway 8 cm from Right side of dr	46	PD20R	East		2	14	Pa	Cu	1
Doorway Pressure 5 cm (+in)	48	-5	5	5 cm from floor, doorway CL	47	PD5C	East		2	15	Pa	Cu	1
Doorway Temperature 79 cm Left BB	32	-5	79	79 cm from floor, doorway 8 cm from Left side of dr	48	TD79LBB	East		2	16	°C	TC	100
Doorway Temperature 79 cm CL BB	48	-5	79	79 cm from floor, doorway CL	49	TD79CBB	East		2	17	°C	TC	100
Doorway Temperature 79 cm Right BB	64	-5	79	79 cm from floor, doorway 8 cm from Right side of dr	50	TD79RBB	East		2	18	°C	TC	100
Doorway Temperature 70 cm Left Asp	32	-5	70	70 cm from floor, doorway 8 cm from Left side of dr	51	TD70LA	East		2	19	°C	TC	100
Doorway Temperature 70 cm Left BB	32	-5	70	70 cm from floor, doorway 8 cm from Left side of dr	52	TD70LBB	East		2	20	°C	TC	100
Doorway Temperature 70 cm CL Asp	48	-5	70	70 cm from floor, doorway CL	53	TD70CA	East		2	21	°C	TC	100
Doorway Temperature 70 cm CL BB	48	-5	70	70 cm from floor, doorway CL	54	TD70CBB	East		2	22	°C	TC	100
Doorway Temperature 30 cm Left Asp	32	-5	30	30 cm from floor, doorway 8 cm from Left side of dr	55	TD30LA	East		2	23	°C	TC	100
Doorway Temperature 30 cm Left BB	32	-5	30	30 cm from floor, doorway 8 cm from Left side of dr	56	TD30LBB	East		2	24	°C	TC	100
Doorway Temperature 60 cm CL BB	48	-5	60	60 cm from floor, doorway CL	57	TD60CBB	East		2	25	°C	TC	100
Doorway Temperature 50 cm CL Asp	48	-5	50	50 cm from floor, doorway CL	58	TD50CA	East		2	26	°C	TC	100
Doorway Temperature 50 cm CL BB	48	-5	50	50 cm from floor, doorway CL	59	TD50CBB	East		2	27	°C	TC	100
Doorway Temperature 40 cm CL BB	48	-5	40	40 cm from floor, doorway CL	60	TD40CBB	East		2	28	°C	TC	100
Doorway Temperature 30 cm CL Asp	48	-5	30	30 cm from floor, doorway CL	61	TD30CA	East		2	29	°C	TC	100
Doorway Temperature 30 cm CL BB	48	-5	30	30 cm from floor, doorway CL	62	TD30CBB	East		2	30	°C	TC	100

LFL MIDAS Hookup Sheet Instrument and Channel Description					Series:	Underventilated RSE				Revision Date:	3/29/2006		
Measurements	Location Description	z z0=floor	y y0=inside front (door) wall	x x0=inside left wall (looking at door)	Overall Channel Number	Abbr.	MIDAS Station	Module: Channels	Mod.	Mod. Ch. No.	Conv.Units	Wire	Gain
Main Channels	RSEmax=(95,142,98)												
Doorway Temperature 20 cm Left BB	20 cm from floor, doorway 8 cm from Left side of dr	20	-5	32	63	TD20LBB	East		2	31	°C	TC	100
Doorway Temperature 20 cm CL BB	20 cm from floor, doorway CL	20	-5	48	64	TD20CBB	East	4-0-31	4	0	°C	TC	100
Doorway Temperature 20 cm Right BB	20 cm from floor, doorway 8 cm from Right side of dr	20	-5	64	65	TD20RBB	East		4	1	°C	TC	100
Doorway Temperature 5 cm CL BB	5 cm from floor, doorway CL	5	-5	48	66	TD5CBB	East		4	2	°C	TC	100
Rear Temperature at Sampling Location Asp	10 cm from C, 29 cm from Left, 29 cm from Rear	88	113	29	67	TRSampA	East		4	3	°C	TC	100
Rear Temperature at Sampling Location BB	10 cm from C, 29 cm from Left, 29 cm from Rear	88	113	29	68	TRSampBB	East		4	4	°C	TC	100
Rear Temperature low Asp	24 cm from Floor, 20 cm from Right, 20 cm from Rear	24	122	75	69	TR24A	East		4	5	°C	TC	100
Rear Temperature low BB	24 cm from Floor, 20 cm from Right, 20 cm from Rear	24	122	75	70	TR24BB	East		4	6	°C	TC	100
Rear Temperature high Asp	80 cm from Floor, 20 cm from Right, 20 cm from Rear	80	122	75	71	TR80A	East		4	7	°C	TC	100
Rear Temperature high BB	80 cm from Floor, 20 cm from Right, 20 cm from Rear	80	122	75	72	TR80B	East		4	8	°C	TC	100
Front Temperature at Sampling Location Asp	10 cm from C, 29 cm from Left, 10 cm from Front	88	10	29	73	TFSampA	East		4	9	°C	TC	100
Front Temperature at Sampling Location BB	10 cm from C, 29 cm from Left, 10 cm from Front	88	10	29	74	TFSampBB	East		4	10	°C	TC	100
Front Temperature low Asp	24 cm from Floor, 20 cm from Right, 20 cm from Front	24	20	75	75	TF24A	East		4	11	°C	TC	100
Front Temperature low BB	24 cm from Floor, 20 cm from Right, 20 cm from Front	24	20	75	76	TF24BB	East		4	12	°C	TC	100
Front Temperature high Asp	80 cm from Floor, 20 cm from Right, 20 cm from Front	80	20	75	77	TF80A	East		4	13	°C	TC	100
Front Temperature high BB	80 cm from Floor, 20 cm from Right, 20 cm from Front	80	20	75	78	TF80BB	East		4	14	°C	TC	100
Gas Burner Temperature Center		15	71	48	79	TNGasC	East		4	15	°C	TC	100
Gas Burner Temperature Front		15	65	48	80	TNGasF	East		4	16	°C	TC	100
Gas Burner Temperature Rear		15	77	48	81	TNGasR	East		4	17	°C	TC	100
Liquid Burner Temperature 1 cm	1 cm from burner bottom	9	71	48	82	TLiq1	East		4	18	°C	TC	100
Liquid Burner Temperature 2 cm	2 cm from burner bottom	10	71	48	83	TLiq2	East		4	19	°C	TC	100
Liquid Burner Temperature 3 cm	3 cm from burner bottom	11	71	48	84	TLiq3	East		4	20	°C	TC	100
Liquid Burner Temperature 4 cm	4 cm from burner bottom	12	71	48	85	TLiq4	East		4	21	°C	TC	100
Liquid Burner Temperature 5 cm	5 cm from burner bottom	13	71	48	86	TLiq5	East		4	22	°C	TC	100
Liquid Burner Coolant Temperature Inlet	outside RSE				87	TCoolIn	East		4	23	°C	TC	100
Liquid Burner Coolant Temperature Outlet	outside RSE				88	TCoolOut	East		4	24	°C	TC	100
Temperature of Total Heat Flux Gauge Front (SN=131836)	In floor on room CL, 35 cm from Front	0	35	48	89	THFF	East		4	25	°C	TC	100
Temperature of Total Heat Flux Gauge Rear (SN=131835)	In floor on room CL, 35 cm from Rear	0	106	48	90	THFR	East		4	26	°C	TC	100
Tamb	At MIDAS East station				91	Tamb	East		4	27	°C	TC	100
Gravimetric Soot Probe 3 Filter Temperature -front	outside RSE				92	TGravSoot3	East		4	28	°C	TC	100
Gravimetric Soot Probe 4 Filter Temperature-front	outside RSE				93	TGravSoot4	East		4	29	°C	TC	100
Gravimetric Soot Probe 3 Solenoid Signal-front	outside RSE				94	VGravSoot3	East		4	30	V	Cu	1
Gravimetric Soot Probe 4 Solenoid Signal-front	outside RSE				95	VGravSoot4	East		4	31	V	Cu	1
Surface Temperature Back Wall Exterior Upper Left	10 cm down, 20 cm from left (facing at doorway)	207	144?	20?	96	TWallUL	East	3-0-6	3	0	°C	TC	100
Surface Temperature Back Wall Exterior Upper Right	10 cm down, 20 cm from right (facing in doorway)	76?	144?	76?	97	TWallUR	East		3	1	°C	TC	100
Surface Temperature Back Wall Exterior Lower Left	10 cm up, 20 cm from left (facing in doorway)	207	144?	207	98	TWallL	East		3	2	°C	TC	100
Surface Temperature Back Wall Exterior Lower Right	10 cm up, 20 cm from right (facing in doorway)	76?	144?	76?	99	TWallLR	East		3	3	°C	TC	100
Surface Temperature Inside Floor	near front HF gauge	0	35?	49?	100	TFloor	East		3	4	°C	TC	100
Surface Temperature Inside Ceiling	near front sampling probe @ ceiling	98?	11?	31?	101	TCeil	East		3	5	°C	TC	100
Liquid Fuel Temperature @ Flowmeter	outside RSE				102	TAmbFuel	East		3	6	°C	TC	100
Created Channels													
Event Marker 1					103	Event1							
Event Marker 2					104	Event2							
Doorway Velocity 79 cm Left	79 cm from floor, doorway 8 cm from Left side of dr	79	-5	32	105	VD79L			B	0	m/s		
Doorway Velocity 79 cm CL	79 cm from floor, doorway CL	79	-5	48	106	VD79C			B	1	m/s		
Doorway Velocity 79 cm Right	79 cm from floor, doorway 8 cm from Right side of dr	79	-5	64	107	VD79R			B	2	m/s		
Doorway Velocity 60 cm	60 cm from floor, doorway CL	60	-5	48	108	VD60C			B	3	m/s		
Doorway Velocity 40 cm	40 cm from floor, doorway CL	40	-5	48	109	VD40C			B	4	m/s		
Doorway Velocity 20 cm Left	20 cm from floor, doorway 8 cm from Left side of dr	20	-5	32	110	VD20L			B	5	m/s		
Doorway Velocity 20 cm CL	20 cm from floor, doorway CL	20	-5	48	111	VD20C			B	6	m/s		
Doorway Velocity 20 cm Right	20 cm from floor, doorway 8 cm from Right side of dr	20	-5	64	112	VD20R			B	7	m/s		
Doorway Velocity 5 cm	5 cm from floor, doorway CL	5	-5	48	113	VD5C			B	8	m/s		
Soot Volume Fraction from Optical Measurement	6 cm from Soffit, doorway CL	75	-5	48	114	Fv			C	0	n/a		

LFL MIDAS Hookup Sheet Instrument and Channel Description   Revision Date: 8/14/2006

Series: Underventilated RSE

Measurements	x x0=inside left wall (looking at door)	y y0=inside front (door) wall	z z0=floor	Location Description	Overall Channel Number	Abbr.	MIDAS Station	Module: Channels	Mod.	Mod. Ch. No.	Conv. Units	Wire	Gain
Main Channels	RSEmax=(95,142,98)												
O2 at Rear Sampling location	29	113	88	10 cm from C, 29 cm from Left, 29 cm from Rear	0	O2Rear	East	1:0-31	1	0	Vol Fr	Cu	1
Bad channel					1	Bad			1	1			1
CO2 at Rear Sampling location	29	113	88	10 cm from C, 29 cm from Left, 29 cm from Rear	2	CO2Rear	East		1	2	Vol Fr	Cu	1
CO at Rear Sampling location	29	113	88	10 cm from C, 29 cm from Left, 29 cm from Rear	3	CORear	East		1	3	Vol Fr	Cu	1
UH at Rear Sampling location	29	113	88	10 cm from C, 29 cm from Left, 29 cm from Rear	4	UHRear	East		1	4	Vol Fr	Cu	1
Rear Gas Rack Dewpoint				outside RSE	5	DPRear	East		1	5	°C	Cu	1
Rear Sampling location gas temperature 1-near probe				outside RSE	6	TGasRear1	East		1	6	°C	TC	100
Rear Sampling location gas temperature 2-near UH				outside RSE	7	TGasRear2	East		1	7	°C	TC	100
Rear Sampling location water inlet temperature				outside RSE	8	TH2OInRear	East		1	8	°C	TC	100
Rear Sampling location water outlet temperature				outside RSE	9	TH2OOutRear	East		1	9	°C	TC	100
O2 at Front Sampling location	29	10	88	10 cm from C, 29 cm from Left, 10 cm from Front	10	O2Front	East		1	10	Vol Fr	Cu	1
CO2 at Front Sampling location	29	10	88	10 cm from C, 29 cm from Left, 10 cm from Front	11	CO2Front	East		1	11	Vol Fr	Cu	1
CO at Front Sampling location	29	10	88	10 cm from C, 29 cm from Left, 10 cm from Front	12	COFront	East		1	12	Vol Fr	Cu	1
UH at Front Sampling location	29	10	88	10 cm from C, 29 cm from Left, 10 cm from Front	13	UHFront	East		1	13	Vol Fr	Cu	1
Front Gas Rack Dewpoint				outside RSE	14	DPFront	East		1	14	°C	Cu	1
Front Sampling location gas temperature 1- near probe				outside RSE	15	TGasFront1	East		1	15	°C	TC	100
Front Sampling location gas temperature 2- near UH				outside RSE	16	TGasFront2	East		1	16	°C	TC	100
Front Sampling location water inlet temperature				outside RSE	17	TH2OInFront	East		1	17	°C	TC	100
Front Sampling location water outlet temperature				outside RSE	18	TH2OOutFront	East		1	18	°C	TC	100
Liquid Burner Temperature 1 cm	48	71	9	1 cm from burner bottom	19	TLiq1	East		1	19	°C	TC	100
Liquid Burner Temperature 2 cm	48	71	10	2 cm from burner bottom	20	TLiq2	East		1	20	°C	TC	100
Liquid Burner Temperature 3 cm	48	71	11	3 cm from burner bottom	21	TLiq3	East		1	21	°C	TC	100
Liquid Burner Temperature 4 cm	48	71	12	4 cm from burner bottom	22	TLiq4	East		1	22	°C	TC	100
Liquid Burner Temperature 5 cm	48	71	13	5 cm from burner bottom	23	TLiq5	East		1	23	°C	TC	100
Liquid Burner Coolant Temperature Inlet				outside RSE	24	TCoolIn	East		1	24	°C	TC	100
Liquid Burner Coolant Temperature Outlet				outside RSE	25	TCoolOut	East		1	25	°C	TC	100
Temperature in space under floor				under RSE	26	TAmbFuel	East		1	26	°C	TC	100
Tamb				At MIDAS East station	27	Tamb	East		1	27	°C	TC	100
Optical Soot (U/fo in adj file)	48	35	75	6 cm from Soffit, doorway CL	28	OptSoot	East		1	28	V	Cu	1
Optical Soot Laser Reference Signal	48	106	75	6 cm from Soffit, doorway CL	29	OptSootRef	East		1	29	V	Cu	1
Optical Soot Apparatus Temperature				outside RSE	30	TOptSoot	East		1	30	°C	TC	100
Gravimetric Soot Probe 1 Filter Temperature - rear			79	outside RSE	31	TGravSoot1	East	2:0-31	1	31	°C	TC	100
Gravimetric Soot Probe 2 Filter Temperature -rear				outside RSE	32	TGravSoot2	East		2	0	°C	TC	100
Gravimetric Soot Probe 1 Solenoid Signal -rear	64		79	outside RSE	33	VGravSoot1	East		2	1	V	Cu	1
Gravimetric Soot Probe 2 Solenoid Signal -rear				outside RSE	34	VGravSoot2	East		2	2	V	Cu	1
Gravimetric Soot Probe 1 Mass Flow -rear				outside RSE	35	SootMF1	East		2	3	V	Cu	1
Gravimetric Soot Probe 2 Mass Flow-front				outside RSE	36	SootMF2	East		2	4	V	Cu	1
Total Heat Flux Gauge Front (SN=131836)	48	35	0	In floor on room CL, 35 cm from Front	37	HFF	East		2	5	kW/m2	Cu	100
Total Heat Flux Gauge Rear (SN=131835)	48	106	0	In floor on room CL, 35 cm from Rear	38	HFR	East		2	6	kW/m2	Cu	100
Doorway Pressure 79 cm Left (+out)	32	-5	79	79 cm from floor, doorway 8 cm from Left side of dr	39	PD79L	East		2	7	Pa	Cu	1
Doorway Pressure 79 cm CL (+out)	48	-5	79	79 cm from floor, doorway CL	40	PD79C	East		2	8	Pa	Cu	1
Doorway Pressure 79 cm Right (+out)	64	-5	79	79 cm from floor, doorway 8 cm from Right side of dr	41	PD79R	East		2	9	Pa	Cu	1
Doorway Pressure 60 cm (+out)	48	-5	60	60 cm from floor, doorway CL	42	PD60C	East		2	10	Pa	Cu	1
Doorway Pressure 40 cm (+in)	48	-5	40	40 cm from floor, doorway CL	43	PD40C	East		2	11	Pa	Cu	1
Doorway Pressure 20 cm Left (+in)	32	-5	20	20 cm from floor, doorway 8 cm from Left side of dr	44	PD20L	East		2	12	Pa	Cu	1
Doorway Pressure 20 cm CL (+in)	48	-5	20	20 cm from floor, doorway CL	45	PD20C	East		2	13	Pa	Cu	1
Doorway Pressure 20 cm Right (+in)	64	-5	20	20 cm from floor, doorway 8 cm from Right side of dr	46	PD20R	East		2	14	Pa	Cu	1
Doorway Pressure 5 cm (+in)	48	-5	5	5 cm from floor, doorway CL	47	PD5C	East		2	15	Pa	Cu	1
Doorway Temperature 79 cm Left BB	32	-5	79	79 cm from floor, doorway 8 cm from Left side of dr	48	TD79LBB	East		2	16	°C	TC	100
Doorway Temperature 79 cm CL BB	48	-5	79	79 cm from floor, doorway CL	49	TD79CBB	East		2	17	°C	TC	100
Doorway Temperature 79 cm Right BB	64	-5	79	79 cm from floor, doorway 8 cm from Right side of dr	50	TD79RBB	East		2	18	°C	TC	100

LFL MIDAS Hookup Sheet Instrument and Channel Description — Series: Underventilated RSE — Revision Date: 8/14/2006

Measurements	Location Description	x x0=inside left wall (looking at door)	y y0=inside front (door) wall	z z0=floor	Overall Channel Number	Abbr.	MIDAS Station	Module: Channels	Mod.	Mod. Ch. No.	Conv.Units	Wire	Gain
Main Channels		RSEmax=(95,142,98)											
Doorway Temperature 70 cm Left Asp	70 cm from floor, doorway 8 cm from Left side of dr	32	-5	70	51	TD70LA	East		2	19	°C	TC	100
Doorway Temperature 70 cm Left BB	70 cm from floor, doorway 8 cm from Left side of dr	32	-5	70	52	TD70LBB	East		2	20	°C	TC	100
Doorway Temperature 70 cm CL Asp	70 cm from floor, doorway CL	48	-5	70	53	TD70CA	East		2	21	°C	TC	100
Doorway Temperature 70 cm CL BB	70 cm from floor, doorway CL	48	-5	70	54	TD70CBB	East		2	22	°C	TC	100
Doorway Temperature 30 cm Left Asp	30 cm from floor, doorway 8 cm from Left side of dr	32	-5	30	55	TD30LA	East		2	23	°C	TC	100
Doorway Temperature 30 cm Left BB	30 cm from floor, doorway 8 cm from Left side of dr	32	-5	30	56	TD30LBB	East		2	24	°C	TC	100
Doorway Temperature 60 cm CL BB	60 cm from floor, doorway CL	48	-5	60	57	TD60CBB	East		2	25	°C	TC	100
Doorway Temperature 50 cm CL Asp	50 cm from floor, doorway CL	48	-5	50	58	TD50CA	East		2	26	°C	TC	100
Doorway Temperature 50 cm CL BB	50 cm from floor, doorway CL	48	-5	50	59	TD50CBB	East		2	27	°C	TC	100
Doorway Temperature 40 cm CL BB	40 cm from floor, doorway CL	48	-5	40	60	TD40CBB	East		2	28	°C	TC	100
Doorway Temperature 30 cm CL Asp	30 cm from floor, doorway CL	48	-5	30	61	TD30CA	East		2	29	°C	TC	100
Doorway Temperature 30 cm CL BB	30 cm from floor, doorway CL	48	-5	30	62	TD30CBB	East		2	30	°C	TC	100
Doorway Temperature 20 cm Left BB	20 cm from floor, doorway 8 cm from Left side of dr	32	-5	20	63	TD20LBB	East		2	31	°C	TC	100
Doorway Temperature 20 cm CL BB	20 cm from floor, doorway CL	48	-5	20	64	TD20CBB	East	4 to 0-20	4	0	°C	TC	100
Doorway Temperature 20 cm Right BB	20 cm from floor, doorway 8 cm from Right side of dr	64	-5	20	65	TD20RBB	East		4	1	°C	TC	100
Doorway Temperature 5 cm CL BB	5 cm from floor, doorway CL	48	-5	5	66	TD5CBB	East		4	2	°C	TC	100
Rear Temperature at Sampling Location Asp	10 cm from C, 29 cm from Left, 29 cm from Rear	29	113	88	67	TRSampA	East		4	3	°C	TC	100
Rear Temperature low Asp	24 cm from Floor, 20 cm from Right, 20 cm from Rear	75	122	24	68	TR24A	East		4	4	°C	TC	100
Rear Temperature high Asp	80 cm from Floor, 20 cm from Right, 20 cm from Rear	75	122	80	69	TR80A	East		4	5	°C	TC	100
Front Temperature at Sampling Location Asp	10 cm from C, 29 cm from Left, 10 cm from Front	29	10	88	70	TFSampA	East		4	6	°C	TC	100
Front Temperature low Asp	24 cm from Floor, 20 cm from Right, 20 cm from Front	75	20	24	71	TF24A	East		4	7	°C	TC	100
Front Temperature high Asp	80 cm from Floor, 20 cm from Right, 20 cm from Front	75	20	80	72	TF80A	East		4	8	°C	TC	100
Temperature of Total Heat Flux Gauge Front (SN=131836)	In floor on room CL, 35 cm from Front	48	35	0	73	THFF	East		4	9	°C	TC	100
Temperature of Total Heat Flux Gauge Rear (SN=131835)	In floor on room CL, 35 cm from Rear	48	106	0	74	THFR	East		4	10	°C	TC	100
Gravimetric Soot Probe 3 Filter Temperature -front	outside RSE				75	TGravSoot3	East		4	11	°C	TC	100
Gravimetric Soot Probe 4 Filter Temperature-front	outside RSE				76	TGravSoot4	East		4	12	°C	TC	100
Gravimetric Soot Probe 3 Solenoid Signal-front	outside RSE				77	VGravSoot3	East		4	13	V	Cu	1
Gravimetric Soot Probe 4 Solenoid Signal-front	outside RSE				78	VGravSoot4	East		4	14	V	Cu	1
Surface Temperature Back Wall Exterior Upper Left	10 cm down, 20 cm from left (facing in doorway)	20?	144?	877	79	TWallUL	East		4	15	°C	TC	100
Surface Temperature Back Wall Exterior Upper Right	10 cm down, 20 cm from right (facing in doorway)	76?	144?	877	80	TWallUR	East		4	16	°C	TC	100
Surface Temperature Back Wall Exterior Lower Left	10 cm up, 20 cm from left (facing in doorway)	20?	144?	10?	81	TWallLL	East		4	17	°C	TC	100
Surface Temperature Back Wall Exterior Lower Right	10 cm up, 20 cm from right (facing in doorway)	76?	144?	10?	82	TWallLR	East		4	18	°C	TC	100
Surface Temperature Inside Floor Front	near front HF gauge	49?	35?	0	83	TFloorF	East		4	19	°C	TC	100
Surface Temperature Inside Ceiling	near front sampling probe @ ceiling	31?	11?	98?	84	TCeil	East		4	20	°C	TC	100
Surface Temperature Inside Floor Rear	near rear HF gauge				85	TFloorR	East		4	21	°C	TC	100
Created Channels													
Event Marker 1					86	Event1							
Event Marker 2					87	Event2							
Doorway Velocity 79 cm Left	79 cm from floor, doorway 8 cm from Left side of dr	32	-5	79	88	VD79L			B	0	m/s		
Doorway Velocity 79 cm CL	79 cm from floor, doorway CL	48	-5	79	89	VD79C			B	1	m/s		
Doorway Velocity 79 cm Right	79 cm from floor, doorway 8 cm from Right side of dr	64	-5	79	90	VD79R			B	2	m/s		
Doorway Velocity 60 cm	60 cm from floor, doorway CL	48	-5	60	91	VD60C			B	3	m/s		
Doorway Velocity 40 cm	40 cm from floor, doorway CL	48	-5	40	92	VD40C			B	4	m/s		
Doorway Velocity 20 cm Left	20 cm from floor, doorway 8 cm from Left side of dr	32	-5	20	93	VD20L			B	5	m/s		
Doorway Velocity 20 cm CL	20 cm from floor, doorway CL	48	-5	20	94	VD20C			B	6	m/s		
Doorway Velocity 20 cm Right	20 cm from floor, doorway 8 cm from Right side of dr	64	-5	20	95	VD20R			B	7	m/s		
Doorway Velocity 5 cm	5 cm from floor, doorway CL	48	-5	5	96	VD5C			B	8	m/s		
Soot Volume Fraction from Optical Measurement	6 cm from Soffit, doorway CL	48	-5	75	97	Fv			C	0	n/a		

145

Description of data columns (channels) in reduced ASCII data files.

Channel Name (units)	Position (cm) RSEmax=(95,142,98)			Description of Measurement
1 Time From Ignition (s)	NA	NA	NA	Time relative to ignition defined by Event 1
2 O2Rear (mol/mol)	29	113	88	Rear O2 volume fraction corrected for water (wet)
3 CO2Rear (mol/mol)	29	113	88	Rear CO2 volume fraction corrected for water (wet)
4 CORear (mol/mol)	29	113	88	Rear CO volume fraction corrected for water (wet)
5 THCRear (mol/mol)	29	113	88	Rear Total Hydrocarbons volume fraction corrected for water (wet)
6 SootRear (g/g)	29	113	88	Rear Soot Mass fraction corrected for water (wet)
7 O2Front (mol/mol)	29	10	88	Front O2 volume fraction corrected for water (wet)
8 CO2Front (mol/mol)	29	10	88	Front CO2 volume fraction corrected for water (wet)
9 COFront (mol/mol)	29	10	88	Front CO volume fraction corrected for water (wet)
10 THCFront (mol/mol)	29	10	88	Front Total Hydrocarbons volume fraction corrected for water (wet)
11 SootFront (g/g)	29	10	88	Front Soot Mass fraction corrected for water (wet)
12 HFR (kW/m2)	48	106	0	Total Heat Flux at Rear Floor
13 HFF (kW/m2)	48	35	0	Total Heat Flux at Front Floor
14 TambCal (C)	NA	NA	NA	Ambient Temperature in LFL
15 TRSampA (C)	29	113	88	Aspirated thermocouple at rear sample location
16 TFSampA (C)	29	10	88	Aspirated thermocouple at front sample location
17 TR24A (C)	75	122	24	Aspirated thermocouple in RSE
18 TR80A (C)	75	122	80	Aspirated thermocouple in RSE
19 TF24A (C)	75	20	24	Aspirated thermocouple in RSE
20 TF80A (C)	75	20	80	Aspirated thermocouple in RSE
21 TFloorR (C)	49	106	0	Bare-bead thermocouple near heat flux gauge
22 TFloorF (C)	49	35	0	Bare-bead thermocouple near heat flux gauge
23 TCeilF (C)	31	11	98	Bare-bead thermocouple on ceiling
24 TWallUR (C)	19	144	87	Bare-bead thermocouple on outside rear wall
25 TUFloor (C)				Bare-bead thermocouple in space below floor
26 TD79LBB (C)	32	-5	79	Bare-bead thermocouple in doorway
27 TD79CBB (C)	48	-5	79	Bare-bead thermocouple in doorway
28 TD79RBB (C)	64	-5	79	Bare-bead thermocouple in doorway
29 TD70LA (C)	32	-5	70	Aspirated thermocouple in doorway
30 TD70LBB (C)	32	-5	70	Bare-bead thermocouple in doorway
31 TD70CA (C)	48	-5	70	Aspirated thermocouple in doorway
32 TD70CBB (C)	48	-5	70	Bare-bead thermocouple in doorway
33 TD30LA (C)	32	-5	30	Aspirated thermocouple in doorway
34 TD30LBB (C)	32	-5	30	Bare-bead thermocouple in doorway
35 TD60CBB (C)	48	-5	60	Bare-bead thermocouple in doorway
36 TD50CA (C)	48	-5	50	Aspirated thermocouple in doorway
37 TD50CBB (C)	48	-5	50	Bare-bead thermocouple in doorway
38 TD40CBB (C)	48	-5	40	Bare-bead thermocouple in doorway
39 TD30CA (C)	48	-5	30	Aspirated thermocouple in doorway
40 TD30CBB (C)	48	-5	30	Bare-bead thermocouple in doorway
41 TD20LBB (C)	32	-5	20	Bare-bead thermocouple in doorway
42 TD20CBB (C)	48	-5	20	Bare-bead thermocouple in doorway
43 TD20RBB (C)	64	-5	20	Bare-bead thermocouple in doorway
44 TD5CBB (C)	48	-5	5	Bare-bead thermocouple in doorway
45 VD79L (m/s)	32	-5	79	Bi-directional probe velocity in doorway
46 VD79C (m/s)	48	-5	79	Bi-directional probe velocity in doorway
47 VD79R (m/s)	64	-5	79	Bi-directional probe velocity in doorway
48 VD60C (m/s)	48	-5	60	Bi-directional probe velocity in doorway
49 VD40C (m/s)	48	-5	40	Bi-directional probe velocity in doorway
50 VD20L (m/s)	32	-5	20	Bi-directional probe velocity in doorway
51 VD20C (m/s)	48	-5	20	Bi-directional probe velocity in doorway
52 VD20R (m/s)	64	-5	20	Bi-directional probe velocity in doorway
53 VD5C (m/s)	48	-5	5	Bi-directional probe velocity in doorway
54 HRRcal (kW)				Heat Release Rate from Calorimeter
55 HRRburner (kW)				Heat Release Rate from Burner (gas, pool or spray)
56 StackMFR (kg/s)				Exhaust hood mass flow rate
57 Tstack (C)				Exhaust hood temperature (near bi-directional probe)
58 O2stack (mol/mol)				Exhaust O2 volume fraction (dry)
59 CO2stack (mol/mol)				Exhaust CO2 volume fraction (dry)
60 COstack (mol/mol)				Exhaust CO volume fraction (dry)
61 THCstack (mol/mol)				Exhaust total hydrocarbons volume fraction (dry)
62 MSstack (mg/m3)				Exhaust soot mass concentration (wet)
63 H2ORear (mol/mol)	29	113	88	Rear water volume fraction (calculated from CO/CO2)
64 H2OFront (mol/mol)	29	10	88	Front water volume fraction (calculated from CO/CO2)

D. Equipment List

Description	Manufacturer	Model	Serial#	NIST#
Oxygen analyzer for HRR	Servomex	540A		549709
CO_2/CO analyzer for HRR	Seimens	Ultramat 6		615207
Total HC analyzer for HRR	Rosemount	400A		569041
Mass flow controller for HRR	MKS	1179A53C	000346712	
Dew Point Transmitter for HRR	Vaisala	DMT242	A4850006	
Sample dryer for HRR	PermaPure	PD-200T-72SS	973-0905-6	
Micro GC for natural gas	Agilent	3000A		623489
Sample pump for HRR	Gast	MOA-P122-AA	4Z026	
Liquid fuel turbine flow meter	Exact Flow			
Natural gas flow meter	Instromet	IRMA 15M-125	319396	605032
Total heat flux gauge (HF Front)	Medtherm	16-0.75-10-4-12-36-20679k	131836	
Total heat flux gauge (HF Rear)	Medtherm	16-0.75-10-4-12-36-20679k	131835	
Oxygen Analyzer (O2Rear)	Servomex	4100	393063	623487
Oxygen Analyzer (O2Front)	Servomex	4100	393064	623488
CO_2/CO Analyzer (Rear)	Seimens	Ultramat 6E	NI-L00197	600671
CO_2/CO Analyzer (Front)	Seimens	Ultramat 6E		609425
Total HC analyzer (Rear)	Baseline-Mocon	8800 H		625764
Total HC analyzer (Front)	Baseline-Mocon	8800 H		623892
Dew point meter (Rear)	Vaisala	DMT242	B074008	
Dew point meter (Front)	Vaisala	DMT242	B074009	
Mass flow controller (Rear&Front)	MKS	M100B53C		
MFC power supply	MKS	247D		
Gas Chromatograph (Front)	HP	5890A	2843A20868	541824
GC column	Restek	Rt-QPLOT	19718	
Pressure Transducer for Velocity	MKS	220DD		
Flow meter (spot check flows)	Bios Dry Cal	DCLT 20K		
Sample pump for aspirated TCs and gas sample tests #1-6	Gast	DOA-P703-FB		
Gas conditioning system (Rear&Front)	PermaPure	MG-2812		rental
Glass-lined stainless steel tubing	Grace Davison	3149		
Soot sample MFC	MKS	M100B53CCS1BV	021407828	
Soot sample MFC	MKS	M100B53CCS1BV	021407829	
MFC power supply	MKS	247D	000763015	
Soot sample filter	Pall	P5PJ047		
Soot sample cleaning pad	Hoppe's	1203		
Soot sample filter holder	Gelman Sciences	2220		
Soot sample 3-way solenoid valves	Parker	04F30C2208AAF4C05		
Soot sample pumps	Gast	MOA-P122-AA